*f*P

OTHER BOOKS BY GERALD SCHROEDER

Genesis and the Big Bang
The Discovery of Harmony Between Modern Science and the Bible

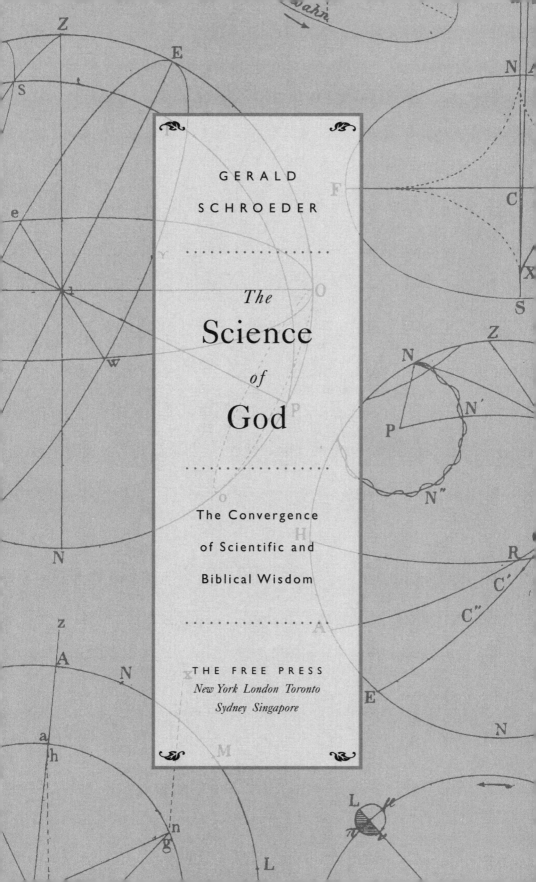

GERALD
SCHROEDER

The
Science
of
God

The Convergence
of Scientific and
Biblical Wisdom

THE FREE PRESS
New York London Toronto
Sydney Singapore

THE FREE PRESS
A Division of Simon & Schuster, Inc.
1230 Avenue of the Americas
New York, N.Y. 10020

Designed by Pei Loi Koay

Manufactured in the United States of America

10   9   8   7   6   5   4   3   2   1

Library of Congress Cataloging-in-Publication Data

Schroeder, Gerald L.
    The science of God : the convergence of scientific and
  biblical wisdom / Gerald Schroeder
        p.   cm.
    Includes bibliographical references and index.
    ISBN 0–684–83736–6
    1. Bible and science.   I. Title.
  BS650.S35   1997
  261.5'5—dc21   97-14978
                                                       CIP

To my wife and children,

who asked the many hard questions,

and to those rare scientists and theologians

who admit that the questions exist.

The heavens speak of the Creator's glory and the sky proclaims God's handiwork.

—PSALMS 19:2

The only path to knowing God is through the study of science—and for that reason the Bible opens with a description of the creation.

—MAIMONIDES, *GUIDE FOR THE PERPLEXED* (1190)

CONTENTS

. . . .

LIST OF ILLUSTRATIONS

. . . .

ふ

A Colleague
Turns
Sixty

The day a friend and colleague turned sixty, I was fortunate to share with him the wait for a bus ride home. "Until I turned sixty," he said, "I never realized how little time I had left." In the years that followed, I watched his frantic race trying to discover why he'd been doing what he'd been doing for the past sixty years.

Not why he and I had spent decades using high-tech physics to fine-tune low-tech farming. In several regions of the developing world we had been able to double farm yields with little or no additional capital investment. The reason for our efforts was obvious: starvation is not pretty anywhere.

His question was much more basic. Why bother being "good"? Is there a transcendent aspect of life that warrants our being good, that might give a meaning to our lives that is fundamentally different from that of other animals?

For someone who waits until age sixty to ask the meaning of life, what the ultimate in life can be, the awakening can be frightening.

In 1894, Albert Michelson delivered the main address at the dedication of the Ryerson Physical Laboratory of the University of

Chicago. Michelson took the opportunity to declare that "The more important fundamental laws and facts of physical science have all been discovered." The physics community thought that there was not much new to learn about our universe.

Ten years later Albert Einstein published the first of his papers that were to revolutionize our understanding of nature and the universe. Einstein's discovery that energy and matter are actually two different forms of the same entity ($E = mc^2$), that matter can be made from energy, and that the flow of time is not a constant, changed mankind's paradigm of the world. His work rested on research performed by Albert Michelson.

Changing one's paradigm is not easy.

Millennia passed before humankind discovered that energy is the basis of matter. It may take a few more years before we prove that wisdom and knowledge are the basis of, and can actually create, energy which in turn creates matter.[1]

"Two things fill the mind with ever increasing wonder and awe—the starry heavens above me and the moral law within me" (Immanuel Kant, 1724–1804).

I propose that they are one and the same whispering voice.

CHAPTER I

Has Science

Replaced

the

Bible?

The

Great Debate

What about dinosaurs?

If the Bible is true, why doesn't it mention dinosaurs? I've been asked that question hundreds of times in places as far-flung as Jerusalem, Los Angeles, Adelaide, and Capetown. It seems to be the universal (or more modestly, the global) biblical perplexity.

Dinosaurs, of course, are a foil for a more basic question: Has science replaced the Bible as the ultimate source of truth?

Nietzsche claimed the discoveries of Copernicus, Galileo, and Darwin had laid God and the Bible definitively to rest. In the spring of 1966, *Time* magazine wondered if he might have been correct, asking on its cover, "Is God Dead?"

Nietzsche's argument is hardly new. Voltaire thought it humorous that this putative God of the Bible would be interested in the affairs of life within the thin film of biosphere that coats the Earth. Voltaire misperceived the biblical concept of an infinite Creator, not to mention the significance of our biosphere. Compared to infinity, are the 10^{27} grams that make up our Earth significantly smaller than the 10^{56} grams of the entire universe? But for him, Copernicus was enough. We aren't the center of the universe, so the Bible must be misconceived.

Misconceptions are what the great debate is all about.

Today, universities have science classes galore on all phases of the mechanics of the universe, from black holes to bacteria. Unfortunately, scientific investigation stops at an account of how the universe functions. It cannot go further. The attempt to discern if a purpose to existence underlies the how is left as a private exercise, one that is usually neglected.

And so the quest that underlies the question of dinosaurs remains. It is a topic guaranteed to draw a full house.

What keeps this great debate alive is that both sides of the theological aisle have an abundance of facts upon which to draw. Unfortunately, in its zeal to protect imagined biblical turf, the church has often claimed more for God's assumed interventions in nature and less for God's laws of nature than the Bible itself claims.★

The thought that religion and science must be at odds is ill conceived. Current surveys consistently report that in Western countries most people (in excess of 70 percent) believe in some form of evolution and in a Divine Creator.[1] Yet within this belief there lies the misperception that religion and science form a dichotomy rather than a duality: there is scientific truth and there is spiritual truth. And the two arise from intrinsically distinct sources, knowledge and intellect giving rise to the former; faith providing the basis for the latter.

Despite all the Bible–science confrontations, despite the battles over high-school textbooks and controversies about government codes on how and when to teach evolution, the fact is science and religion are both thriving.

For centuries, the meanings of various passages in the Bible have been disputed. Some interpretations have been hostile to science, others to the text itself. This erosion of biblical understanding is tragic, and we've paid a price for it. We don't need a temple priest or a university philosopher to measure the decay in the fabric of our

★I use the term Church as a generality for organized religion in the West. No specific denomination is necessarily intended.

society. Our 60 percent divorce rate and the double locks on our doors provide a succinct summary of the effect.

This book accepts neither Bible nor science as being individually sufficient for a hungry mind seeking explanations of and purpose in life. In that sense, it is for skeptics and religious believers alike. These seemingly disparate sources of knowledge are combined as a single data base from which generalized conclusions are drawn. What appear to be diametrically opposed biblical and scientific descriptions of the creation of the universe, of the start of life on Earth, and of our human origins are actually identical realities but viewed from vastly different perspectives. Once these perspectives are identified, they coexist comfortably with all the rigorous science and traditional belief anyone could demand.

The medieval philosopher Moses Maimonides wrote that conflicts between science and the Bible arise from either a lack of scientific knowledge or a defective understanding of the Bible. This is a continuing problem. Acknowledged experts in science may assume that although scientific research requires diligent intellectual effort, biblical wisdom can be attained through a simple reading of the Bible. Conversely, theologians who have devoted decades to plumbing the depths of biblical wisdom often satisfy their scientific curiosity through articles in the popular press and then assume they can evaluate the validity of scientific discoveries. The "opposition" is viewed with a level of knowledge frozen at a high school or pre–high school level. No wonder the "other side" seems superficial, even naive. To relate these two fields in a meaningful way requires an in-depth understanding of both. Nobel laureate and high energy physicist Steven Weinberg is unsympathetic to the idea that ancient commentators on the Bible foresaw modern cosmological concepts regarding the origin of our universe. Yet in his recent book *Dreams of a Final Theory,* he readily admits, "It should be apparent that in discussing these things . . . I leave behind any claim to special expertise."

For the religious believer, it is time to render unto Einstein that which is Einstein's: science has given us a powerful tradition for the examination of life as we know it. Scientists are not always right, but they are very good about testing their own theories and correcting their mistakes. Their discoveries daily reveal wonders in the workings of our universe. The idea that scientific explanation of nature's marvels detracts from the grandeur of creation is both absurd and ill-conceived. When understood in context, this knowledge can be a source of inspiration.

The late Professor Richard Feynman, formulator of much of modern physics, in the opening volume of his classic, *The Feynman Lectures on Physics,* describes what science can do for religion:

> *Poets say science takes away from the beauty of the stars—mere globs of gas atoms. Nothing is "mere." I too can see the stars on a desert night, and feel them. But do I see less or more? The vastness of the heavens stretches my imagination—stuck on this carousel my little eye can catch one-million-year-old light. A vast pattern—of which I am part—perhaps my stuff was belched from some forgotten star, as one is belching there. Or see them with the greater eye of Palomar, rushing all apart from some common starting point when they were perhaps all together. What is the pattern, or the meaning or the why? It does not do harm to the mystery to know a little about it. For far more marvelous is the truth than any artists of the past imagined! Why do the poets of the present not speak of it? What men are poets who can speak of Jupiter if he were like a man, but if he is an immense spinning sphere of methane and ammonia be silent?*[2]

But science has its limitations and the skeptic too must realize this. It can never speak to the purpose of life, the "why" which Feynman so emphasized. Colleagues who follow in the footsteps of Einstein would do well to render unto the Bible that which is the Bible's, the search for purpose.

Using logic, scientific knowledge, and ancient biblical interpretations, I discuss here the duality, not the dichotomy, of science and Bible.

As a scientist, I look at the universe and try to extract underlying principles by which it functions. I rely on the inherent consistency of nature. If the laws of nature are not fixed, if they are being tampered with in some miraculous way, then science is useless. The consistency of nature is a basic tenet of all scientific inquiry.

The consistency of nature is also a basic tenet of biblical religion. Eight hundred years ago, the kabalist Nahmanides wrote that "since the world came into existence, God's blessing did not create something new from nothing, instead the world functions according to its natural pattern."[3] In secular terms, kabalah had stated that the laws of nature were and are adequate to channel our universe toward the development and sustenance of life.

Professor Weinberg is an avowed skeptic, if I understand him correctly, but even he agrees with Nahmanides. "Life as we know it," Weinberg writes, "would be impossible if any one of several physical quantities had slightly different values. . . . One constant does seem to require incredible fine tuning."[4] This constant has to do with the energy of the big bang. Weinberg quantifies the tuning as one part in 10^{120}. Scientific notation is an understatement and so I will expand that exponential into decimal notation. If the energy of the big bang were different by one part out of

100000000000000000000000000000000000000000000000 000000000000000000000000000000000000000000000000 0000000000000000000000000

there would be no life anywhere in our universe. The universe is tuned for life from its inception. Genesis agrees: when life first appears on the third day, the word creation does not appear. We are merely told "The earth brought forth" life. Earth had within it the necessary properties for life to flourish.

Michael Turner, the widely quoted astrophysicist at the University of Chicago and Fermilab, described that tuning with a simile: "The precision," he said, "is as if one could throw a dart across the entire universe and hit a bullseye one millimeter in diameter on the other side."

Scientists discover astonishing facts every day. Whether it's physicists realizing how a tiny difference in temperature at the time of the big bang would have obviated the possibility of life as we know it, or biochemists discovering the miraculously complex, delicately balanced molecular machinery that makes blood clotting possible, scientists face the wonders of our existence very directly.[5] Why, then, is it that a hard core of vocal scientists are avowed atheists, taking their discoveries of the wonders of nature as sufficiently edifying ends in themselves?

My wife, the author Barbara Sofer, was privy to the notes of a private meeting held at Princeton between the late prime minister of Israel David Ben-Gurion and Albert Einstein. Their conversation might have turned toward the politics of the young State of Israel. Instead, immediately they focused on what really intrigued them, whether there was evidence for a higher force directing the universe. Both agreed there was such a force, a central power. Yet neither Ben-Gurion nor Einstein had a feeling for formal religion.

The spark is there in all of us. Still, we may intellectually reject the very explanations our emotions tell us are true.

Aristotle, 2,300 years ago, observing that nothing comes from nothing, assumed that nothing ever will—or did. Therefore he defined the universe as eternal. This stood in sharp contrast to the claim made a thousand years earlier in the opening sentence of the Bible, that there had been a beginning to the universe: "In the beginning . . ." (Gen. 1:1). Aristotle found no conflict between an eternal universe and belief in a supernatural god. He believed in a host of them.

The Bible's claim for a creation and Aristotle's denial of it gave impetus to several early attempts at estimating the age of the universe. The most quoted of these estimates is the much maligned calculation made by James Ussher, archbishop of Armagh, Ireland (1581–1656).

Summing the generations listed in the Hebrew Bible and then estimating the reigns of rulers thereafter, he arrived at an autumn creation date: high noon, 23 October 4004 B.C.E. The exaggerated exactness seems a bit bizarre. But then how would a cleric know about creations of universes? Not surprisingly, a contemporary of Ussher's, the famous German astronomer Johannes Kepler (1571–1630), the scientist who discovered that the planets revolve around the Sun in elliptical, not circular, orbits, disagreed with Ussher's estimate. Kepler thought the creation occurred in the spring!

Today, their use of the Bible for the purpose of science seems misplaced. What does the Bible have to do with scientific reality?

One's turf is spiritual, the other's physical, or so we have been taught. Actually, the Bible, properly understood, can be a handmaiden of science (and vice versa). As such it is instructive to note that Ussher's and Kepler's calculations of an approximately six-thousand-year-old universe are infinitely closer to our current estimate of time since the big bang than was either Aristotle's opinion or that of two thirds of the leading U.S. astronomers and physicists, who in a 1959 survey agreed with Aristotle. Human logic sided with Aristotle but was in error. The biblical paradigm of a beginning to our universe, a creation, was correct. The error in the biblical age of the universe was not in the Bible, but in how Ussher and Kepler used the details of the Bible to make their calculations.

Though Kepler was a committed Christian, his work had the touch of heresy. (The idea that any discovery can be heretical is beyond me, but the Church managed to do away with quite a few scientific heretics. Galileo escaped only because of his longtime friendship with the pope.) Kepler's discovery of the elliptical orbit of the planets did not sit well with the religious establishment. Circles were perfect geometric shapes, ellipses defective. An infinitely powerful God would be expected to produce perfect orbits. The Bible did not claim this. The Church did.

Isaac Newton (1642–1727), a devout Christian, applied laws of gravity and inertial motion to planetary motion. These properties, first conceptualized by Galileo Galilei, state that a body continues in its present state of motion or rest unless acted upon by an outside force such as gravity or friction. It must have come to him as a lightning bolt out of the blue to find himself accused by none other than the renowned mathematician Gottfried Leibnitz, co-inventor of the calculus, of bringing "occult qualities and miracles into philosophy." Leibnitz felt gravity was "subversive of revealed religion." What occult qualities? What subversion? According to Leibnitz, with inertial motion, the planets could keep up their motion without God's hand continuously pushing them along. Of course the Bible makes no such claim of God's constant planetary push. The laws of nature, as part of the creation package, were and are adequate for the job. Is there a theologian alive today who believes gravity subverts the grandeur of the biblical God?!

If I had to assign chief blame for the ongoing struggle between science and religion and the resulting erosion of biblical credibility, it would be to the leaders of organized religion. Since Nicolaus Copernicus (1473–1543) had the audacity to suggest that the Sun, not Earth, was the center of our solar system, their knee-jerk reaction to any scientific discovery that impinges on our cosmic origins has been to deny its validity. Only later, sometimes centuries later, do they bother to gather the facts.

Copernicus was a believing Catholic as well as a prominent astronomer. His discovery did not shake his faith. What does the position of the Earth have to do with belief in a creator of the universe or the validity of the Bible? Nowhere does the text claim that Earth is central to anything. In fact, the very first sentence of the Bible—

"In the beginning God created the heavens and the earth" (Gen. 1:1)—places the heavens before Earth.

But overly enthusiastic clerics, staking some imagined claim for biblical truth, extended Genesis to include what it never did: a positioning of the Earth.

As scientific data demonstrating the Sun's centrality accumulated, the Church was forced into embarrassed retreat. The popular perception is that science had proven the Bible wrong. In reality, the claim of Earth's centrality had nothing to do with the Bible.

A century passed before theologians reluctantly adjusted to Newton's laws of motion and a universe that did not revolve around Earth. Then, as if adding insult to injury, Charles Darwin appeared on the scene with the *Origin of Species* and a claim for evolution. The year was 1859.

The thought that life in general and humans in particular had developed from lower life forms through random mutations was simply unacceptable to the Church. (We have discovered during the past three decades that it is also substantially unacceptable to science, but that is a topic for later chapters.) The concept of evolution was condemned as heretical, notwithstanding the fact that Darwin in the closing lines of his book attributed the entire evolutionary flow of life to "its several powers having been originally breathed by the Creator in a few [life] forms or into one." Nonetheless, the gauntlet of heresy had been thrown down. Darwin's adherents, though not Darwin himself, readily took it up.

Thomas Henry Huxley (1825–1895), known as Darwin's bulldog, wasted no time. In 1860, just one year after publication of *Origin,* he attacked with a vengeance. An idea attributed to, but not found in, the opening chapter of Genesis purported to show that each species was a special creation unto itself. This was absolutely contrary to Darwin's concept of the gradual evolution of species. For Huxley the scant fossil record, which today in its richness brings so much of Darwin's theory into doubt, was absolute proof that Darwin had guessed correctly. In Huxley's words, "History has embalmed for us [as fossils] the speculations upon the origin of living beings."

Huxley must have been aware that Darwin did not base his theory on the fossil record. Darwin realized that the staccato nature of the fossil record in no way confirmed evolution via natural selection. Rather, Darwin noted the morphological changes produced by breeders of pigeons and other domesticated animals, and assumed (quite likely in error) that if in tens of generations lean ancestral stock evolve into robust productive progeny, then gradually over tens of millions of generations vastly greater changes would have occurred, changes so great that phylum by phylum life rose ever higher on the imagined evolutionary tree.

Such an evolutionary tree has yet to be discovered in the fossil record. But to Huxley the gaps in the fossil record were no obstacle. He had his preconceived notions and the facts were not going to stand in his way. And so he wrote, "The myths of paganism [read here the Hebrew Bible] are as dead as Zeus and the man who should revive them in opposition to the knowledge of our time would be justly laughed to scorn. . . . In the 19th century, the cosmogony of the semi-barbarous Hebrew is the opprobrium of the orthodox. . . . The doctrine of special creation [of each species] owes its existence very largely to the supposed necessity of making science accord with the Hebrew cosmogony."[6] This is the same Huxley who later promoted a falsified fossil record that purportedly proved the smooth evolution of the modern horse.

I doubt that the author of the Hebrew Bible was either pagan or barbarous. The text is far too clever for a barbarian author. As for pagan, the basics of western society find their origins in the Five Books of Moses. Huxley wouldn't have been Huxley without them.

Furthermore, Huxley's reliance on the fossil record to eventually prove Darwin's thesis of gradual evolution is now known to be misplaced. The statement Darwin repeats several times in *Origin of Species,* "natura non facit saltum"—that nature does not make jumps—is simply false. Transitional forms are totally absent from the fossil record at the basic level of phylum and rare if present at all in class. Only after basic body plans are well established are fossil transitions observed. Darwin would have been much closer to the truth had he written "natura solum facit saltum"—that nature only makes jumps. In the words of Niles Eldredge, curator at the American Museum of Natural History in New York City, "The fossil record we were told to find for the past 120 years [since Darwin] does not exist."[7]

Unfortunately, though the crude nature of Huxley's attack on the Bible was ill-placed, his argument with the superficial understanding of the opening chapters of Genesis was in order. There is no biblical suggestion that each species had a separate creation, a claim that is so much anathema to avowed evolutionists. In fact, during the main discussion of land animals on the last of the six days of creation, the word creation does not even appear.

Here we come to a basic tension between religion and science: biblical literalism. Haven't those who demand a literal reading of Genesis noticed that Genesis is literally filled with contradictions? How can such a strange and poetic text be read literally? Two millennia ago, long before paleontologists discovered fossils of dinosaurs and cavemen, long before data from the Hubble and Keck telescopes hinted at a multibillion-year-old universe, the Talmud stated explicitly that the opening chapter of Genesis, all thirty-one verses, is presented in a manner that conceals information.[8] The kabalistic tradition has come to elucidate that which is held within those verses. Kabalah is logic, not mysticism, but logic so deep that it might seem mystical to the uninitiated. Literalism is simply not an effective way to extract meaning from the Bible.

Consider this example. The account of each of the first three days of the creation week closes with ". . . and there was evening and there was morning . . ." (Gen. 1:5, 8, 13). Nothing unusual about that until we arrive at the fourth day to discover that only now does the author produce a sun (Gen. 1:14–16). Having evening and morning on the first three days without a sun might have encouraged the adult reader to look beyond a simple reading of the text (as we do later).

Here's another: Adam is told "Of every tree of the garden you may eat freely. But of the tree of the knowledge of good and evil you shall not eat, for on the day that you eat of it you shall surely die" (Gen. 2:16, 17). The verb in the Hebrew text is doubled to emphasize the certainty of the punishment for transgression, hence "surely die." So what does Adam do? As typically human, he eats of it. And then he lives another 930 years (Gen. 5:5). Did I miss something? I thought he was to die on the day he ate from the tree.

Has the author, divine or otherwise, forgotten what was written a few passages before? With a literal reading, the second sentence of the Bible contradicts the first sentence. Whether we see the Bible as the direct word of God, divinely inspired, or totally of human origin, its subtleties and pathos have kept the Western world's interest for ages. The author was smart. These contradictions are not by chance and not errors. They are beacons urging us to seek the deeper meanings held within the text just as we seek meanings within the subtleties of nature.

The first step in a rapprochement between science and Bible is for each camp to understand the other. Distancing the Bible from a few misplaced theological shibboleths will do wonders in furthering this mutual understanding.

I have already treated several. Earth need not be at the center of the universe for biblical religion to survive. As Genesis 1:1 stated, first came the heavens and then came Earth. Western religion has learned to forego its misplaced dream of a universe revolving around Earth, to accept gravity as a part of nature and not the machinations of a perverted mind, and most important, to read the Bible, as Moses insisted three times on the day of his death, as a poem, as a text having within it a subtext harboring multiple meanings (Deut. 31:19, 30; 32:44).

The mistaken shouts of protest against the imagined heresy of gravity have faded to distant echoes. As later chapters argue, the same will happen with the more recent theological cries directed against evolution. The biblical account of animal life's development, which amounts to a mere eight verses (!), will have no problem with the final scientific understanding of how animal life evolved. What is needed on both sides is patience, not the diatribes of a T. H. Huxley or the sophistry of biblical literalism.

The scientific concept of evolution has already come to embrace what Darwin himself, and a century later John Maynard Smith and now Stephen Jay Gould, insist upon: a flow of life channeled by laws inherent to the universe.[9,10] The level at which this channeling dominates and where it gives way to uncertain meanderings contingent upon local factors remains to be determined. But a channel, confining the biology of evolution to a limited range, is obvious even to avowedly secular scientists.

Approximately 250 million years ago, 95 percent of all marine species suffered a massive extinction.[11,12] The ecology was wide open for innovation, yet no new body plans evolved to fill the ecological space. Why? Almost four billion years ago, an exquisite, efficient system for encoding and transmitting the information needed to guide an organism's development from seed to adult appeared. That same system, the double helix of our genetic DNA, to this day guides all forms of life, from algae to oak trees, from microscopic bacteria to massive elephants and humans as well.[13] Is only one genetic system workable? Can only a few body plans satisfy the laws of nature? Based on all biological and paleontological data, that seems to be the case. But why?

These constraints are not by chance. They reveal a limit, a definition, in fact a channel, for the breadth of choices available to the development of life. Discoveries in molecular biology and paleontology deepen the channel almost daily. (For example, the same gene has been discovered to control the development of all visual systems in all phyla. Again only one option seems to be viable; a topic for consideration in later chapters.) The great schism between science and religion which has characterized the past five hundred years may at last be narrowing.

Obviously, the biblical concept of an infinitely powerful Creator demands that in this infinity, It can produce and control all of life at will. But there is not a hint in the Bible that this control is constantly exercised. Instead, to quote Nahmanides once more, "the world [channeled by the laws of nature] functions according to its natural pattern." Consider just three episodes of the many that make this so clear.

1. To aid in the conquest of Canaan, God promises to send hornets in order to make the enemy flee (Deut. 7:20). Here's God controlling nature. Just two verses later (Deut. 7:22), we read "And the Lord thy God will cast out those nations little by little [why little by little?] . . . lest the beasts of the field increase upon you." Notice the problem? If God can control the hornets to drive out the nations, why doesn't God also keep the beasts from multiplying? Nature is given free rein at this level.

2. Of the twelve tribes of Jacob, only the tribe of Levi was to serve directly in the Temple. For this they must be physically fit. The Bible provides a list of birth defects which disqualify a Levi from fulfilling this potential (Lev. 21:17–23). Why have birth defects? The biblical concept of an infinite God is a God that could make all births perfect. I imagine if I were God I would. But the world as described in the Bible does not function according to our demands. Most children are born healthy and physically normal, but not all. Nature has its level of freedom.

3. "And God saw the light, that it is good" (Gen. 1:4); "And God saw that it [the oceans and earth] is good" (Gen. 1:10); "And God saw that it [the origin of plant life] is good" (Gen. 1:12); and on and on.* God sees that "it is good" seven times in the thirty-one verses of the first chapter of Genesis—the creation chapter. Almost a quarter of all those verses are devoted to God's discovering that "It is good." Didn't God realize from the start that it would be good? Perhaps. But this is not explicit in the text.

Time and again, the Torah implies that the infinitely powerful biblical God withheld control and allowed the world to follow its own course. With this godly approach to world management, the results were not always "good." The Creator then redirected the flow.

Adam and Eve are placed "in the Garden of Eden to work it and to keep it." No complaint about the work—it seems Adam did not expect a free lunch. "And the Eternal God commanded the man, saying, From every tree in the garden you shall surely eat." The one

---

*"It was good" is a mistranslation of the Hebrew text that satisfies the English but misses the cosmic intent.

request was that "from the tree of the knowledge of good and evil you shall not eat" (Gen. 2:15–17). It was all too tempting. Adam and Eve ate and God expelled them from Eden. The rest is history.

Adam and Eve had two children, Cain and Abel. In the biblical account of human life to this point, these four are it. Cain murdered Abel (Gen. 4:8). That doesn't say much for producing a society steeped in brotherly love. God exiled Cain. Adam and Eve restarted the process with their third son, Seth.

Perhaps the most encompassing Divine retuning of all was the Flood: "And the Eternal saw the wickedness of man was great upon the land. . . . And the Eternal repented that He had made man upon the land. . . . And the Eternal said I will wipe out man whom I have created from the face of the soil, from man to beast to creeping animals and to winged animals of the heavens, for I repent that I have made them" (Gen. 6:5–7).

What can you say after that?

At each stage, God withheld control to a greater or lesser extent. This allowed the world to develop according to the laws of nature created at the beginning and the moral responsibility implanted by the human soul (animals being amoral). A limited experiment was underway. If it failed, Divine retuning (a flood for example) redirected humanity. When it worked, God was pleased: "It's good." This aspect of nature had achieved its divine purpose (Nahmanides on Gen. 1:4, 10, 31).

The eighteenth-century kabalist, Moshe Chayim Luzzatto, was intrigued by the Godly lack of universal control. He framed his book, *The Knowing Heart* (ca. 1734), as a debate between the worldly intellect and the ethereal *neshama* (the human soul). The intellect, being a product of nature (Gen. 1:26), understands that God (Elokim) works through the coordinates and constraints of time, space, and matter. Those are the ways of all existence in our universe. But those very constraints allow for deviations from what might otherwise be perfection. The *neshama,* being a direct creation (Gen. 1:27), is confused. Why bother with the deviations, she (the *neshama*) asks him (the intellect)? Why not just lay down a perfected world having the Eternal totally in control? Through their debate, the paradox of an infinite Creator imposing less than infinite control over the products of creation is resolved. The laws of nature provide direction, but within that direction there is leeway, meanderings contingent upon the immediate environment, just as a river's meanderings are contingent on the local terrain. Though occasionally it may leave in its path an isolated bow lake, the flow eventually reaches the sea. These excursions in the flow of events might be seen as the vicissitudes inherent in an evolutionary process having within it a general direction. In humans, these meanderings are called free will.

The Bible documents one evolutionary change in a physical trait, the trait of longevity (Genesis 5 and 11). The biblical data record a transition that might just as well have come from a modern text on animal husbandry and breeding.

Prior to the flood at the time of Noah, the life spans of the persons being discussed ranged from 365 years to 969 years, with the average being 840 years. Sexual maturity (the age at which a woman first gives birth) was reached at 65 to 187 years (average 115 years). Both averages are approximately ten times the current values for developed countries, obviously far from today's reality. Whatever one may think of the pre-Noah longevity, by the time of Abraham, just ten generations after Noah, life span had so decreased that the Bible required an explicit miracle for Abraham, age 99, and Sarah, age 89, to conceive a child (to be named Isaac, from the Hebrew word for laugh, as Abraham did when the angel said he and Sarah would be parents the following year; Gen. 17:17).

The cause of this dramatic decrease in life expectancy is not stated. However, the actual age data as listed in the Bible are instructive (see Figure 1). Prior to Noah there is no strong trend either increasing or decreasing longevity. Following Noah, a trend is clear. Life span becomes shorter through the generations. The biblical concept is that change takes place over time and through generations, just as did the development of the world in the first chapter of Genesis.

The trend of shortening life span and more rapid sexual maturity is similar to that observed in domesticated animals. After generations of breeding, broilers now reach slaughter size in thirty days instead of three to six months, and beef cattle in about a year instead of two. Both Maimonides in the twelfth century and Nahmanides in the thirteenth suggest that changes in the environment following the

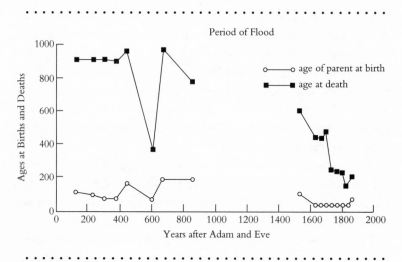

FIGURE I: Ages of births and deaths of persons following Adam and Eve (Genesis 2–11)

flood favored ("selected for" in modern terms) shorter life span. The scatter in the pre-Flood data, both in sexual maturity and longevity, reveal that a range existed from which the shorter spans could be selected. Today this type of selection forms a basis for breeding and population genetics.

The gradual evolution of a trait that only slightly alters the morphology of the animal is referred to as micro-evolution. The change in longevity for post-Flood humans is micro-evolution. It is observed regularly in farmyards and biology laboratories. It finds no dispute in the Bible. Macro-evolution, the evolution of one body plan into another—a worm or insect or mollusk evolving into a fish, for example—finds no support in the fossil record, in the lab, or in the Bible.

So how are we to understand creationism? Biblically, creation is a divine act of *tsimtsum,* contraction—a spiritual contraction by which the Creator *removes* part of Its infinite unity ("Hear Israel the Eternal our God the Eternal is One," Deut. 6:4). Complexity now appears where there had been the undifferentiated simplicity of One. The greater the *tsimtsum,* the more extensive the complexity and the greater the corresponding potential for imperfection.

Isaiah in two sentences clarifies this concept: "I am the Eternal, there is nothing else. I make light and create darkness, I make peace and create evil" (Is. 45:6,7). The infinite source of light withdraws and darkness is created. The infinite source of peace (*shalom,* from the root *shalaim* meaning whole, complete) withdraws and evil (lack of perfection) is created. The first biblical *tsimtsum* (Gen. 1:1) allowed the physical complexity of the universe with its laws of nature to emerge. Then followed the creation of the *nefesh*—the soul of animal life—allowing animals choice strongly dictated by instinct and inclination (Gen. 1:21). The third and final creation was the human soul—the *neshama*—instilling free will in humans (Gen. 1:27).

We humans choose within constraints contingent upon our surroundings. The meanderings of nature and of society produce challenges to each person contingent upon her or his locale. How we react to those challenges provides them with spiritual significance. The moral choices of a German in 1996 are easier than those of a German in 1936. Though man cannot control his environment or even his destiny, his conduct is altogether in his hands.[14]

With each act of *tsimtsum,* the Bible tells us, the channel through which all nature flows broadened. Its license to meander increased.

The conflict between science and the Bible is ironic. Throughout the Bible, knowledge of God is compared with the wonders of nature. As stated so well in Psalms (19:2): "The heavens tell of God's glory and the sky declares his handiwork."

Eight hundred years ago, the medieval philosopher Maimonides wrote that science is not only the surest path to knowing God, it is the only path, and for that reason the Bible commences with a description of the Creation.[15] In some communities that thought was sufficient cause to burn his books.

I am not so naive as to claim that current scientific opinion can explain the workings of all events described in the Bible, or that biblical wisdom foresaw all that modern science has discovered. However, in biology, paleontology, cosmology, among a sweep of topics the confluence is remarkable.

Maimonides' claim has proven itself. God is back in the discussions of science, and with good reason.

The perception that religion requires faith alone is a misperception. Religion requires belief and belief is built on knowledge. For knowledge, we live in an opportune era. The discoveries of the past few decades in astronomy, high energy physics, and paleontology have revolutionized the understanding of our cosmic genesis. They have taken us to the threshold of time and the beginnings of life.

We have learned there was a time before which there was neither time nor space nor matter.[16] Discoveries related to the explosive development of life have forced a reevaluation of the process and direction of evolution.[17,18]

Although the popular impression is otherwise, within the professional scientific community, most of us realize that the Bible is not about to be replaced with a formula that can fit on a T-shirt. The quintessential admission of this appeared in an article written by Harvard University professor Stephen Gould: "Science simply cannot adjudicate the issue of God's possible superintendence of nature."[19] Knowing the plumbing of the universe, intricate and awe-inspiring though that plumbing might be, is a far cry form discovering its purpose.

The flow of time and events from the big bang to the appearance of humankind is summarized in the thirty-one verses with which the Bible begins: Genesis 1. These few hundred words describe sixteen billion years of cosmic history, topics about which entire libraries of books have been written. With a superficial reading of Genesis, and certainly with a superficial reading of the text in translation, we haven't a prayer of understanding the details.

But then, superficiality is a loser in all endeavors. If we relied on casual observations of nature, we would still believe that the Sun revolves around the Earth. That is certainly the simple perception we derive, day by day, from sunrise to sunset.

Unlike abstract concepts of faith, biblical religion has a track record that can be evaluated. As Paul Johnson articulated so incisively, the Bible is the earliest identifiable source of the great conceptual discoveries essential for civilization: equality before the law, sanctity of life, dignity of the individual, individual and communal responsibility, peace as an ideal, love as the foundation of justice.[20]

Might it be, as Einstein and Ben-Gurion concluded at their meeting so many decades ago, that a contingency, an ingenious coherence transcending and joining all aspects of existence does pervade the scientific maze we call our universe?

Rather than merely discarding such a premise as rubbish, or embracing it as the logical and obvious truth, the Bible and science both have the identical response: study the data and from a position of knowledge determine the probability of this coherence having happened by an unguided nature.

Let's look at the universe, its cosmic genesis, and see if we can discern hints of a transcendent Creator historically active in the creation. If we do, we can move on and investigate how we might capture the all-too-rare rush of joy sensed when we chance upon the transcendent. Instead of waiting passively for it to happen, imagine being able to have that joy as a permanent partner in life. That would be called getting the most out of life.

In the following chapters, I attempt to avoid the subjective tendency of bending Bible to match science or science to match Bible. To accomplish this goal, I cite only scientific opinions appearing in these leading science journals. The theological sources are primarily restricted to works that predate by centuries the discoveries of modern science. Those will be primarily the Hebrew Bible, the Talmud (a collection of commentaries on the Hebrew Bible redacted in two stages, 300 and 500), and the thirteenth-century kabalist, Nahmanides. Nahmanides is not only the leading kabalistic commentator on the Book of Genesis (and the entire Torah), but also one of the earliest kabalists whose commentary is written in a readily understandable Hebrew.

The Science of God deals with the Book of Genesis, the heritage of all Western religions. Because of the book's nonsectarian nature, it employs the abbreviation B.C.E., for "before the Common Era," instead of B.C. and C.E. for "during the Common Era," instead of A.D.

CHAPTER 2

The New
Convergence:
Science,
Scientists,
and the
Bible

The origins of modern science are rooted in the works of great thinkers who were also religious believers, persons such as Copernicus, Bacon, Galileo, Kepler, Newton, Leibnitz. It is ironic that in part through their work, points of contention between science and religion arose that are neither the artifacts of ignorance nor merely related to particular schools of biblical interpretation. The immanence of God is one. Few scientists can accept the concept of a God that intervenes in our daily lives, choosing at will to violate the laws of nature. The big three questions about origins—of the universe, of life, and of humans—have also divided the secular from the religious. Until the 1960s, most scientists came to believe that the universe had no beginning. Most believers insisted it did. Most scientists also came to believe that life on Earth and mankind arose gradually. Many believers insisted they were sudden acts of creation.

These positions would seem to be irreconcilable. On the one side is a very active God, breaking Its own rules, impatient with a gradual genesis. On the other side is an eternal universe, with mathematical laws at work "on their own." Yet as the detailed workings of the cosmos and of life have unfolded, the separations have narrowed and in significant cases the opinions have merged. The rapprochement began in physics and is slowly spreading to biology. Reconciliation does not require that every scientist become a believer, nor that every believer embrace all aspects of science. It will be complete when we accept the need to read and understand the Bible on the Bible's terms—a text that carries a subtext requiring interpretation—and when scientists, having already discovered that there is a limit to knowledge, admit that science is powerless to confirm or deny a purpose for life.

We will never discover an absolute proof for or against the existence of a Creator by looking through a microscope or a telescope. We can, however, study trends in science and determine if they tend toward confrontation or confluence with the Bible's unchanging worldview.

THE PLUMBING OR THE PURPOSE OF THE UNIVERSE

Roger Penrose, professor of mathematics at Oxford University, has among his areas of expertise the study of the universe shortly after its creation. He was awarded the Wolf Prize for his analytic description of the big bang, which forms the basis of all big bang cosmology.

Penrose finds the laws of nature tuned for life. This balance of nature's laws is so perfect and so unlikely to have occurred by chance that he avers an intelligent "Creator" must have chosen them.[1] It is as if we were written into the equations of the universe at its inception or in the words of physicist Paul Davies, "built into the scheme of things in a very basic way."[2]

Not every student of nature agrees.

Nobel laureate Weinberg also studies the physics of the early universe. Unlike Penrose, Weinberg is saddened by the fact that, although the laws of nature show "incredible fine tuning," from all of his research into the substance and mechanics of the early universe, he finds the universe to be pointless, that life is only "a little above the level of farce."[3]

Studying the plumbing of the universe, how it functions physically, can lead to very different conclusions regarding its purpose. It is all a matter of perspective. If we believe the universe is the natural outcome of gravity and nuclear physics with a quantum fluctuation to get it off the ground, that's the way we may find it. Or we might see it as divine.

THE BEGINNING OF THE UNIVERSE

From the time of Aristotle, 2,300 years ago, scientific theory held the universe to be eternal. The unchanging stellar pattern of the heavens was shining evidence of this eternity. Einstein even claimed to have proven it, though with some sleight-of-hand. Through the early 1960s in the face of mounting evidence to the contrary, two thirds of leading U.S. scientists surveyed believed it.[4] For 3,300 years, since the revelation on Sinai, the Bible denied it, steadfastly claiming there was a beginning to our universe.

Only in the past thirty years has science resolved the question. Based on data from telescopes and particle accelerators, the 3,300-year-old Genesis 1:1 was correct all along. There was a beginning. The old, we have learned, is not necessarily the out of date.

Much of the appeal for the Aristotelian theory of an eternal universe was that a universe with a beginning poses a problem that borders on theology. As Weinberg wrote in the closing page of his classic *The First Three Minutes,* "Some cosmologists are philosophically attracted to the oscillating model [of the universe], especially because, like the steady-state [eternal] model, it nicely avoids the problem of Genesis." The "problem of Genesis" is of course the problem of a beginning. Weinberg, and many scientists since, proceeded to demonstrate on theoretical grounds that the laws of nature demand a beginning for our universe, a time before which there was neither time nor space nor matter.

Because we have lived for the past thirty years with strong evidence for the big bang, we take the fact of a beginning as an obvious given, as popular wisdom. But let us not lose perspective. This shift in scientific opinion, after millennia of opposition, represents the most significant change science can ever make toward biblical philosophy. Evolution, dinosaurs, cavemen are all trivial controversies when compared to the concept of a beginning. While a beginning does not confirm the existence of a Beginner, it does open the way for that possibility. It is small wonder then that Professor John Maddox, the admittedly secular former editor-in-chief of the prestigious journal *Nature,* called the big bang "philosophically unacceptable."[5]

It is to the credit of the scientific method that scientists have been willing to accept data which originally seemed logically unacceptable. Imagination and intuition often provide the basis for a hypothesis. But as data accumulate, imagination must bend before reality. The greatness of science must be that it rests on facts, not on opinion.

Even a scientist as brilliant as Albert Einstein had to learn this lesson.

In 1917, using the laws of general relativity he had published two years earlier, Einstein developed a series of equations describing the condition of the universe. They showed something seemingly illogical, that the universe was dynamic. Vesto Slipher of the Lowell Observatory in Flagstaff, Arizona had already reported astronomical measurements that implied the universe was expanding. Slipher's data rested totally on Einstein's own laws of relativity. But Einstein's mind-set for the concept of an eternal universe was too strong. He rationalized away Slipher's data, subjectively changed his equation so that it described a static universe and published this mutilated form, the famous cosmological equation. As data accumulated indicating that indeed the universe was expanding, Einstein wrote to his colleague and fellow Nobel laureate, Max Born, that his denial of his own theory was the "biggest blunder of my life."

Why the biggest blunder? Einstein realized that if day by day the universe was expanding, getting ever larger, then what about yesterday, a year ago, a millennium ago, and ever backward until billions of years ago there was only a point, a point that marked the beginning. Einstein could have followed his own discoveries and predicted the most important statement ever made relative to man and the universe: there was a creation. And he blew it. He could not give up his opinion in favor of his facts. Desired axioms die hard even in the face of contradictory evidence.

CAUSING A BIG BANG BEFORE TIME

It may come as a surprise that belief in a beginning does not require belief in a Beginner. The laws of nature allow the creation of the universe without the need for a creator. "Quantum uncertainty," an aspect of the physics known as quantum mechanics, allows the small but finite possibility of something coming into being from nothing via what is known as a quantum fluctuation.[6] Since quantum mechanics has a sixty-year track record for predicting and explaining observed phenomena, this theory of a quantum fluctuation/big bang beginning too may be correct.

There are, however, basic problems with the concept. First of all, quantum fluctuations are phenomena that relate to the laws of nature within our universe. The beginning of our universe marks the beginning of time, space, and matter. There is every indication that the big bang also marks the beginning of the laws of nature. If this is true, then prior to the existence of the universe there was no nature and therefore there were no laws of quantum mechanics by which to engender the needed quantum fluctuation.

Alternatively, we might believe that the laws of nature are eternal. That would suppose the existence of laws for a universe that does not exist. Purposeless laws. This stretches one's imagination. If on the other hand we accept these eternal laws to be part of something more grand, we are back to the Bible. Theology claims the laws of nature are indeed eternal. They are contained within the infinite, eternal biblical God. The laws of nature are the few characteristics of this Infinity that are projected in our universe. Time, for example, may be the projection of eternity into a finite material universe.

Even with eternal natural laws, the very idea of a fluctuation causing the creation brings us to the problem of cause and effect. Effects are separated from causes by time. But before the universe, time did not exist. On the fact of time's nonexistence prior to the creation, theology and science are in complete agreement. So the beginning of the universe can not be the effect of a cause. The universe must simply "be." As *Nature*'s editor, John Maddox, summarized so accurately, "It is an effect whose cause cannot be identified or even discussed."[7] His thought was far from original. Both the Talmud[8] and kabalah[9] stated this a millennium and more ago.

And then we confront another problem: a quantum fluctuation in what? Not in empty space, not in time, and not in matter. Those did not exist prior to the universe. There was no vacuous void within which the universe was to appear. On this, science and theology are also in complete agreement. A mathematician colleague, Dennis Turner, posted this question on a newsgroup bulletin board of the worldwide web. The only replies were nonanswers. The most notable was a classic of in-house humor from the noted mathematician/physicist John Baez of MIT: "This is one of those questions they only tell you the answer to when you get your PhD, and they make you promise not to tell anyone."

Finally, a quantum vacuum fluctuation can only produce a universe that is closed, that is, a supermassive universe, one that might have collapsed shortly after its big bang. There are no data that support the contention that our universe is closed.[10,11]

The only "secular" way of surmounting the problem of the beginning is to theorize that there exists an infinitely large, eternal macro-universe within which are embedded an infinite number of finite universes. An analogy would be an infinite turbulent sea having an infinite number of bubbles ever forming and expanding. Each bubble would be a universe, having its own duration, its own laws of nature. Some bubbles bud and give rise to other bubbles. One of the bubbles would be our universe.

To say that there is no observable proof for such a macro-universe is an understatement. The laws of nature exclude the possibility of seeing outside our universe even if there is an outside. It is a theory that can never be tested by observation.

At this point it seems that the theory of an eternal Creator and the theory of an infinite universe both rely on a leap of faith. The Bible (Deut. 4:39) claims it is possible to "know" which leap is in the correct direction—to know not as an absolute proof, but rather "beyond all reasonable doubt." And how can we know? Ironically, time and again, the Bible by analogy shows that the needed knowledge can come from the supposed adversary: the study of nature.

For years I had been on that adversary's team. As a scientist trained at the Massachusetts Institute of Technology I was convinced I had the information to exclude Him—or is it Her?—from the grand scheme of life. But with each step forward in the unfolding mystery of the cosmos, a subtle yet pervading ingenuity, a contingency kept shining through, a contingency that joins all aspects of existence into a coherent unity. While this coherence does not prove the existence of a Designer, it does call out for interpretation.

DESIGN BY CHANCE

The design in nature's move toward life might be pure chance. Perhaps we are here because good fortune smiled and produced by chance the multitude of events needed to coax life from the chaos of the big bang. It would be something like winning a lottery for which a million people—even a billion—had purchased tickets. A miracle? Not really—someone had to win and you were the lucky person.

A bit of scrutiny reveals the shortcomings of this analogy. You see, if you win a lottery this week and then again next week, and then again the third week, chances are that before you collect your third winnings, you will be on your way to jail for having rigged the results. The probability of winning three in a row, or three in a lifetime, is so small as to be negligible. "But," you plead before the judge, "probability never says never. It was just a rare set of circumstances." It's true that probability never says never, but all of physics, which means all of nature, is based on the understanding that the very very very unlikely never happens. Without this basic understanding, there is no foundation for any assumptions of physics or cosmology.

With the universe we did not win just one lottery. We won at the choice for the strength of the electromagnetic force (which encourages atoms to join into molecules). We won at the strength of the strong nuclear force (which holds atomic nuclei together; were it a bit stronger the diproton and not hydrogen would be the major component of the universe, and no hydrogen means no shining stars). Other winning lotteries were the strength of the weak nuclear force and the strength of gravity (which dominates the universe at distances greater than the size of molecules and clusters mass into galaxies, stars, and planets), the mass and energy of the big bang, the temperature of the big bang, the rate of expansion of the universe, and much more. Lottery upon lottery, and all winners. They have meshed to produce the wonderful world in which we live. By chance? Not if our understanding of the laws of nature is even approximately correct. To this observer of nature, our universe looks like a put-up job.

I'm not alone in that opinion. Molecular biologist, and observant Catholic, Professor Michael Behe stated the case for the believing biologist: "You can be a good Catholic and believe in Darwinism [random mutations of genes sifted by nature's selection]. Biochemistry has made it increasingly difficult, however, to be a thoughtful scientist and believe in it."

TUNED FOR LIFE

Consider carbon, element number six in the periodic table. Before it come hydrogen, helium, lithium, beryllium, and boron. After it come nitrogen and oxygen and the rest of the ninety-two natural elements. Life as we know it is based on carbon. It is the only element that can form the long and complex chains necessary for the processes of life. Elsewhere in the universe life may be based on liquids other than water, but carbon is the necessary elemental jack-of-all-trades for life. It is also the essential Lego-like stepping stone for the production of the eighty-six natural elements heavier than carbon.

But the formation of carbon sits on a knife edge of uncertainty. To form carbon, radioactive beryllium (element number 4) must absorb a nucleus of helium (element number 2) and build to element number 6 (carbon). That seems simple enough: $4 + 2 = 6$. (Who said physics is difficult?) However beryllium does not have much of an existence. The mean life of this radioactive beryllium atom is 10^{-16} seconds.

To get the feel for the brevity of 10^{-16} seconds, look at that number in decimal notation: 0.0000000000000001 seconds. In that sliver of time, a helium nucleus must find, collide with, and be absorbed by the beryllium nucleus, thus metamorphosing into the atomic staff of life.

The only way the helium nucleus can be absorbed in the brief time before the beryllium decays is if the energies of these nuclei are matched exactly to the required energies of excitation. And matched they are. Carbon is the fourth most abundant element in the universe and the most abundant element that is solid at temperatures when water is liquid. If this reaction were foiled by mismatch, the universe would contain hydrogen and helium and not much of anything else. The elements of life would not have formed.

We stand in awe at the beauty of diamond white stars stretching across a velvet black desert night sky. That's our emotional response.

These discoveries of science provide a quantifiable basis for that awe. The universe in which we live is very special.

THE BEGINNING OF LIFE

It is not just physics and chemistry that are begging religious questions. Biology is too. Instead of Aristotle, biologists have Darwin to thank for the rupture. And they are still being attacked by anti-Darwin theologians. So on their part it isn't surprising to find hostility. During the past few decades, as molecular biology has moved to the forefront, some of the same elements that motivated physicists to ponder the presence of contingent order in the universe are emerging in biology: wonder at the intricacy of life's cellular workings; the resistance to efforts to remove all possibility of purpose.

Until the 1970s the scientific theory of the origin of life claimed that billions of years passed on the newly cooled Earth during which inorganic elements randomly coupled and broke apart, coupled and broke apart, until finally after a myriad of these random trials, a self-replicating molecule formed that led to primitive life.

The test of this theory was a search for fossils among the most ancient rocks able to bear fossils (sedimentary rocks). To the amazement of the scientific community, fossil evidence was discovered that showed life started, not after the predicted billions of years, but immediately on the cooled Earth. The billions-of-years-to-produce-life theory had to be discarded. It had been the leading and most logical scientific opinion into the 1970s. But how else other than gradually by a multitude of random reactions could life have started? Nobel laureates had waxed poetic in their treatises extolling the wonders of the random beginning of life as it gradually emerged from a primordial inorganic soup. They were wrong.

To account for life's immediate appearance, today's scientific theories require either that life was planted on Earth from outer space (!), or that an exotic property of molecular self-organization rapidly joined the necessary chemicals into self-replicating molecules and then a yet-to-be-discovered series of catalysts developed these fecund molecules into life itself. The most optimistic scenarios reveal that random reactions on their own could not have produced life in the time available even if the entire universe were the laboratory and testing ground for these random reactions.[12-14]

For the 3,300 years of its existence, Genesis has presented to us its account of the origin of life. Liquid water appeared on Earth and life appeared immediately thereafter (both occur on biblical day three). The Bible tells us (correctly) not only the timing of the origin of life (at the appearance of liquid water), but also proposes the mechanism for its origins: "The earth brought forth [life]" (Gen. 1:11). No mention of a special creation is associated with the start of life. The earth itself had the special properties to orchestrate the beginning of life. In modern terms, those properties are described as self-organization and/or catalysts.

The first part of the biblical theory for the origin of life has been proven to be true: life appeared on Earth immediately after water. As such, it may be that the second part, the need for self-organization and catalysts, is also true. There is no biblical indication that life was planted here from outer space.

Science has made the two most important steps it can ever make in closing ranks with the Bible: (1) there was a beginning to our universe, and (2) life started rapidly on Earth and not via millennia of purely random reactions. These are global concepts. The nuances of our origins are equally instructive.

NO GRADUALISM IN EVOLUTION

The development of life is as perplexing as its immediate inception. According to the fossil record, gradual evolution has been found to be false at every major morphological change.

First, one-celled life sprang into being as soon as water was present, 3.8 billion years ago. One might have expected that complex multicellular organisms would then have developed in orderly successive stages. Such was not the case. Instead of a gradual steady thrust of life evolving complex structures, 3.2 billion years passed during which life remained confined to one-celled organisms, followed 650 million years ago by the simple globular forms of uncertain identit, known as Ediacaran fauna.

Then, 530 million years ago in the Cambrian era, with no hint in earlier fossils, the basic anatomies of all life extant today appeared simultaneously in the oceans.[15–18] The Cambrian explosion of life is one of the century's greatest discoveries.

The biblical parallel to this account is striking. Genesis 1 announces the beginning of life on the third day, immediately after the first appearance of liquid water on Earth (also day three). One might have expected that in the biblical rush to get life started, the Bible would then proceed at once with animal life. But such is not the case. There is a lapse, a hiatus. During the entire fourth day life is not mentioned.

Then, on day five, Genesis describes an explosion of life. The waters, we are told, "swarm[ed] abundantly with moving creatures that have life" (Gen. 1:20). In the following chapters I discuss the relativity of time and the fidelity of the biblical cosmic clock to Earth time. Day five, we will discover, corresponds to this Cambrian era, 530 million years ago. In the present overview, I'll confine the discussion to the sequence.

Following this explosion of multicellular life at the start of the Cambrian era, for some puzzling reason, no other new phyla (basic anatomies) ever appeared. Classes of animals developed within each phyla but they always retained the basic body plan of their particular phylum.

Intra-phylum evolution presents a further puzzle. Consistently new organisms, whether among plant groups or animals, make their first fossil appearance highly specialized and fully developed, last their time, and disappear. Rarely if ever are there fossil indications that competition by a new and better-adapted class of animals outstripped an outdated ancestor in the race for food, shelter, and survival, thus driving the ancestor to extinction. The fossil record regularly fails to give any hint at the basic anatomical levels that a change in morphology was in the offing.

The origin of wings is a classic example researched by paleontologists and biologists.[19] There is no clue in the fossil record of their developmental origins. Fossils directly below (i.e., more ancient than) flying insects are wingless insects. Wings emerge suddenly, fully developed and quite large—reaching a 30-centimeter span. The ichthyosaurus marine reptile has a fish-like body. It first appears in the fossil record during the Jurassic era with fully developed fins, paddles, and bill. In excess of 100 million years later, at its extinction, it is essentially the same. In the plant kingdom, 140 million years ago angiosperms literally blossomed forth. There is no inkling in older fossils of their impending explosion.

An accurate description of macro-evolution as presented by the fossil record is that it usually takes place somewhere else and all we are left with is the punctuations. Darwin realized this far better than his overly enthusiastic followers. On no less than seven occasions in the *Origin of Species,* he implored his readers to ignore the evidence of the fossil record as a refutation of his concept of evolution or to "use imagination to fill in its gaps." The record's leaps and bounds, he claimed, were the result of its being incomplete.

Though Darwin's excuse may ring hollow in light of today's extensive paleontological evidence, there are potentially valid reasons why transitional fossil forms might be lacking. The transition might have occurred in a small isolated population. Or the transition species, being not yet fully robust, might have had fewer members than the final morphology. Fewer members can mean fewer fossils. Perhaps viruses, by transferring genes across phylum or class lines, induced sudden changes in the morphology. In that case there would never have existed a transitional stage. Acidic soils, such as are commonly found in forests, may decompose the skeleton before fossilization can occur.

The magnificent Natural History Museum in London devotes an entire wing to demonstrating the fact of evolution. They show how pink daisies can evolve into blue daisies, how gray moths change into black moths, how over a mere few thousand years, a wide variety of cichlid fish species evolved in Lake Victoria. It is all impressive.

Impressive, until you walk out and reflect upon that which they were able to document. Daisies remained daisies, moths remained moths, and cichlid fish remained cichlid fish. These changes are referred to as micro-evolution. In this exhibit, the museum's staff did not demonstrate a single unequivocal case in which life underwent a major gradual morphological change.

The standoff between gradual evolution and the staccato punctuations of the fossil record is reminiscent, almost a déjà vu, of an era in physics just prior to the establishment of quantum mechanics.

Classical physics had told us that as the temperature of a body increases, the spectral radiation emitted from that body should increase as a smooth and continuous function of the temperature. This was the only logical expectation. Then came experimental data showing stepwise, quantized changes. And after much intellectual struggle, quantum mechanics, illogical and counterintuitive though it is, replaced classical physics. Stepwise changes in morphology are also illogical but observed in the fossils. Might they also be an accurate description of the processes that govern the development of life, even if they cannot yet be explained?

Unlike radiation emission, the validity by which the fossil record presents the history of evolution cannot be tested. The closest hints to such a test are the similarities in genetic material (DNA) across divergent forms of life and the ontogeny of embryos. The former is a mixed blessing, with some genes showing great similarity among diverse animals and others—in which similarity is expected—showing little.[20–22] Ontogeny indeed produces what appears to be an evolution. In the human embryo, for example, there is a transition through stages found in embryos of fish, then reptiles, and finally mammals.[23] Though this tells nothing of how these changes were first induced (were they the result of random point mutations of the DNA?) and at what rate they occurred, it is suggestive.

Among professionals active in evolutionary biology, such as Gould and Dawkins, Eldredge and Smith, a battle rages over whether gradual evolution ever occurred and if it did, why it is not evident in the fossils. The ferocity of the battle sometimes suggests that sudden leaps in the record would imply God's direct role in evolution while gradualism would mean randomness and no role for God. This is nonsense. Contrary to popular lay opinion, the Bible is mute concerning the driving mechanism behind macro-evolution. A few basic animal body plans are created on day five and that is the last mention of creation for animals. Six sentences later, part way through day six, the account of the evolution or development of life is complete.

Biblical time before Adam is so highly compressed that there is simply no opportunity to describe the processes (or even the sequence except in the broadest of terms) that caused life to advance from the simple to the complex. The Bible is eager to get on with the story of humankind. From Adam and forward, biblical time is time as we know it, no longer compressed. But less than six thousand years span the period between biblical Adam and the present, not nearly enough time for macro-evolution to occur, even by the most extreme interpretation of punctuated evolution.

Evolutionary biology and biblical theology by their very natures are retrospectives, theories of history. Both are bounded by our incomplete knowledge of history.

Cosmology has come to agree that there was a beginning (Gen. 1:1). Biology has discovered that indeed life on Earth started shortly after the appearance of liquid water (Gen. 1:9–12) and that three billion years later animal life exploded in a burst of aquatic organisms (Gen. 1:20–21) hosting all phyla alive today. Those data relate to the physical aspects of life, but what of the spiritual? Does science leave space for the biblical sine qua non of being human: free will?

The theory of determinism, proposed by Marquis Pierre Simon de Laplace (1740–1827), left no option for choice. Chemistry and physics fixed the outcome of every causative act. Effect was determined totally by the conditions of the world at the moment of cause. Or so the skeptics alleged.

Then came quantum mechanics and the discovery that identical causes do not always produce identical effects. At the most basic level of subatomic matter the world is not governed by cause and effect. A single quantum of light, the photon, striking the eye's retina can cause the retina to initiate a signal to the brain. That means our brain is sensitive to quantum effects. And the outcome of a quantum effect is not controlled totally by the physical world. The unavoidable conclusion is that our mental activity is not totally the function of the physics and chemistry of our brain and body the instant before we think. QM is not only the graveyard of determinism, it has restored the concept of free will to its status as a respectable topic among the intellectually enlightened. The biblically inclined had never questioned it.

Yet old ideas cling even in the face of contradictory evidence. It's a biological fact that the song a sparrow learns in its youth is its song for life. We humans are not so very different.

When data mount ever more convincing arguments against a favored paradigm, all sorts of mental machinations allow us to retain our preconceived notions of reality. If we have spent much of a lifetime attempting to prove the validity of a premise in question, the emotional stakes are high. Cognitive dissonance, humanity's inherent ability to ignore unpleasant facts, helps us in our struggle to retain the error of our ways. Among evolutionists, this was never shown more clearly than in the case of Charles D. Walcott.

Here is a story, a true story, about cognitive dissonance. The account is special because it involves one of the most famous scientists in his field and the intellectual insult we have all suffered because he suffered from this all-too-human foible. Sadly, it is only a magnified example of a phenomenon that occurs every day both in scientific and religious minds.

Charles D. Walcott had finally reached the Burgess Pass. The adjacent valleys were 5,000 feet below. Walcott loved the Canadian Rockies. It is spectacular country. He was on a combined summer holiday plus field trip, packing by horse across the mountains of eastern British Columbia in search of fossils. And for fossils, the ridge connecting Mount Wapta and Mount Field near the Burgess Pass was to become a very special location. The shale rocks over which Walcott was climbing were about to yield the most important fossils ever found. They were to reveal the origins of all modern life.

The story of the Burgess Pass begins some 560 million years earlier when it was part of a tropical sea. Though charted today as a Canadian mountain 8,000 feet above sea level, a location where the first snows come in mid-September and melt only in late July, 560 million years ago the area was a layer of mud below the shallow waters of a tropical continental shelf. A limestone reef marked the border of the shelf. Beyond this, the bank fell sharply to 200 meters.

Shallow water plus tropical sun plus nutrients equals life, and life there was in abundance. The temperate sunbathed waters teemed with a plethora of marine plants and animals. Soil, washed from the adjacent shore, continually refurbished the nutrients demanded by this vibrant community. Occasionally the accumulated weight of mud and detritus increased beyond that which the reef face could support, and the sediment broke loose. The suddenly freed mud raced down the steep bank, its silt-like flow sweeping up all in its path, wrapping the prey in an oxygen-poor shroud.

The fineness of the mud was important. Normally only hard tissues such as bone and teeth survive the burial process intact. Here the silt captured, and the anoxic environment preserved, the soft outer tissues of the trapped plants and animals. The fine silt, infiltrating the animals' bodies, preserved their inner organs. In time, as layers of successive slides increased the overlying pressure, the mud metamorphosed into shale. Fossils, complete with three-dimensional impressions of their soft tissues and organs, had been formed. A unique history of life lay recorded in stone.

But their bed was far from being a tranquil place of rest.

The simple view of Earth's surface as static masses of land amidst vast oceans is a fiction. Closer to the truth is a picture of continent-sized blocks of semi-rigid stone slowly drifting over hotter, more pliable, silicate stone known as the lower mantle of the Earth. The motion is less than a snail's pace: about one and a half centimeters per year.[24] But geology is patient. A centimeter or so a year is enough to have changed the face of the Earth many times over since it formed from a mass of stardust some 4.5 billion years ago.[25] That tropical sea shelf was on the move toward colder climes. It was to become western Canada.

As continental plates move slowly across the Earth's surface, their leading borders buckle, wrinkling much the way newly fallen snow wrinkles in advance of a plow. In time, portions of the borders are pressed under the advancing continent, just as some snow slips under the plow. Those subsided rocks melt and any tale of fossils they once bore is erased. But other parts rise to form the tops of the wrinkles. These we see as the rugged mountains bordering the western coasts of North and South America. It is our good fortune that the Burgess Shale rose to form one of those wrinkles. Its precious cargo of fossils had been preserved.

Walcott was a world-renowned paleontologist and the world's expert on the explosion of multicellular life that occurred in the Cambrian period, 500 to 600 million years ago. Ever on the watch for new fossils, a slab of the Burgess Shale caught his experienced eye. The rock may have borne a telltale clue: parallel lines scraped onto its surface by a glacier's motion. Ten thousand years before, at the close of the last ice age, glaciers originating in the Arctic had skimmed the top off this mountain. Shale, buried for more than 500 million years, now lay exposed.

Using his geologist's hammer, Walcott would have rapped the multilayered slab on its edge. The layers separated and there, held within, was the fine imprint of a crustacean. But this was impossible. The shale was too old to contain a fossil as complex as this specimen. Some 550 million years ago, at the start of the Cambrian, the only life on Earth was the most simple of forms, one-celled bacteria, algae, protozoans, and some pancake-shaped life of uncertain definition known as Ediacaran fossils.[26,27] There was no way evolution could have advanced life from one-celled protozoans to the complexity of this crustacean in the twenty or so million years of the Cambrian. There simply had not been the time for that development. Well into the 1970s, evolutionary theory assumed that in excess of 100 million years were needed for the basic body plans of advanced life to evolve from the simplicity of pre-Cambrian life.[28]

Other shale pieces yielded a variety of equally fantastic animal fossils. Walcott, meticulous as always, recorded their shapes in his diary. During the next decade Walcott collected and shipped between sixty and eighty thousand of these specimens to his institution in Washington, D.C.

That Walcott realized he had made a major discovery is obvious from the vast number of fossils he collected. Representatives of every animal phylum, the basic anatomies of all animals alive today, were present among those half-billion-year-old specimens. These fossils revealed an extraordinary fact.

Eyes and gills, jointed limbs and intestines, sponges and worms and insects and fish, all had appeared simultaneously. There had not been a gradual evolution of simple phyla such as sponges into the more complex phyla of worms and then on to other life forms such as insects. According to these fossils, at the most fundamental level of animal life, the phylum or basic body plan, the dogma of classical

Darwinian evolution that the simple had evolved into the more complex, that invertebrates had evolved into vertebrates over one hundred to two hundred million years, was fantasy, not fact.

Such a challenge to Darwinian evolution had its professional hazards. In those heady years of the ascent of Darwin to near sainthood, no scientist questioned the role of random evolution, and certainly not if that scientist was director of the Smithsonian Institution, the largest organization of museums and curators of its day. You see, Charles Doolittle Walcott was the director of the Smithsonian. And so, following modest disclosures printed in the *Smithsonian Miscellaneous Collections,* a publication of extremely limited circulation, Walcott reburied the fossils, all sixty thousand of them, this time in the drawers of his laboratory. The year was 1909. Eighty years were to pass before their rediscovery.

Walcott was the personal friend of three United States presidents. He was both professionally and politically powerful. Had he wanted to make a splash in the media with his discoveries, he had the voice of the press and the leverage of politics to carry his message.

When in 1953 Stanley Miller announced the results of an experiment that erroneously were interpreted to prove that life had started by chance random chemical reactions, the news swept the globe. But then, life starting by chance favored the endeared paradigm of a world ruled by chaotic, helter-skelter reactions, bumbling along with no plan. Only twenty-five years later did the scientific community admit its error and retract the theory of life having started by random reactions.[29-31]

Was Walcott's act a conspiracy of silence? In some way it must have been, even if his goal was merely to keep the glory or thrill of the discovery for a time when he personally could accomplish the research effort required by his ancient trove. Walcott's administrative responsibilities were demanding. However, considering the importance Walcott ascribed to the Burgess fossils (remember, sixty thousand of them lined the drawers of his laboratory), the director of the largest system of museums in the world could certainly have mustered the budget to hire a brigade of graduate students for the work. Ironically, it was a graduate student, Simon Conway Morris, who eventually played a key role in the interpretation of these fossils.

Professor Stephen Jay Gould of Harvard University says that Walcott's belief in God kept him from realizing the significance of his find.[32] God has taken the rap for a host of human failures, but this charge takes the cake! Walcott believed in God. He also believed in evolution, gradual evolution, as God working through nature, gradually changing the early simple forms of life into modern complexity. The explosion of life recorded in the Burgess Shale fossils contradicted, even confounded, this gradualism.

To maintain his belief in God's gradual methods, Gould claims, Walcott "shoehorned" the fossils into already known categories. It's a weak argument. If Walcott thought Burgess was just a variation of the same old stuff, why his extraordinary effort to collect so many and then to keep them for himself?

Let's be more realistic and more charitable to his intelligence. Let's just ascribe his act to cognitive dissonance, humanity's inherent desire to ignore unpleasant facts.

Gould, a descendant of the priestly tribe of Levi that served in King Solomon's Temple, seems to have a problem with God.[33] In two essays that first appeared in *Natural History* magazine and then were republished in a Gould collection, he quotes the closing lines of Darwin's *Origin of Species:* "There is grandeur in this view of life. . . . Whilst this planet has gone cycling on according to the fixed law of gravity, from so simple a beginning endless forms most beautiful and most wonderful have been, and are being, evolved."[34]

The reference to "the fixed law of gravity" reflects Darwin's belief that his theory of evolution would suffer a similar fate as Newton's laws of planetary motion. At first it would be attacked as being sacrilegious. But eventually it would be seen to fall within a religious paradigm.

It would, but with no thanks to Gould and his misquote. Here is how Darwin, not Gould, closes the *Origin of Species* (including the sixth edition, 1872, the last edition during Darwin's life): "There is grandeur in this view of life, with its several powers, having been originally breathed by the Creator into a few forms or into one; and that, whilst this planet has gone cycling on according to the fixed law of gravity, from so simple a beginning endless forms most beautiful and most wonderful have been, and are being, evolved."

To accomplish his ruse with no hint to the reader, Gould put a period after "life" and capitalized "Whilst," neatly leaving out any hint of God. It seems that Gould has a problem with Darwin as well as with God. Although Gould states "Charles Darwin my hero and role model,"[34] he is desperate to change Darwin's worldview into Gould's.

The Burgess fossils suggest an explosion of new life forms. Evolutionists might prefer an ordered development of this life, while creationists claim an inexplicable sudden appearance. There is cognitive dissonance on both sides. Why should we assume answers to questions like this? The overwhelming weight of evidence tells us something exotic certainly happened to produce the variety of life as we know it on our planet. As to what that was, the jury is still out.

Rediscovery of Walcott's fossils in the mid-1980s changed the concept of evolution. Their effect has been so dramatic that the most widely read science journal in the world, *Scientific American,* in its November 1992 issue was moved to question: "Has the mechanism of evolution altered?" The same reaction to these fossils appeared in the October 1993 issue of *National Geographic* magazine. The science section of the *New York Times* referred to the fossils as demonstrating "revolution more than evolution."[35] And *Time* magazine featured them in a comprehensive and scientifically accurate cover story titled "Evolution's Big Bang."[36]

It is not the mechanism of evolution that has altered. Rather, it is our understanding of the development of life that must be revolutionized and that is a hard task to accomplish.[37] Like the sparrow, we sing the song we learned in our youth.

Authors of high school textbooks and even introductory courses in college biology still ignore these data. In the college classes I teach, I regularly encounter students who are being taught the tale of invertebrates gradually evolving into vertebrates. At $15,000 a year tuition, that's an expensive error. But then the gradual evolution of phyla is so logical and so much more easily explained. The paradigm a teacher learns in youth is hard to displace.

I must clarify an important point. The Burgess fossils do not question the development of classes of life. It is no secret that each individual phylum first appeared as simple aquatic forms and became more complex with the passage of time. The Book of Genesis proclaimed this fact 3,300 years ago: first came aquatic animals, then winged creatures and land animals, then mammals. That's Genesis 1! The Bible knows about development. Humans are the last, not the first, of the animals mentioned in Genesis. It is inter-phylum development that has been proven to be a fantasy.

The charge of cognitive dissonance must also be laid at the doorstep of the religious community. Theologians, rabbis, priests, and pastors alike, knowing no more science than what they read in the popular press, trash carbon-14 and every other piece of fossil-related data. Calling upon bizarre distortions of nature's laws, they try to fit a superficial knowledge of science into conformity with a simple reading of the Bible. Ironically, carbon-14 has little to do with the fossil record and is absolutely irrelevant to fossils of dinosaurs and those of the Burgess Shale.

Interpretation is as essential for understanding Genesis as it is for understanding nature. It is time for both sides to stop the war. Render unto science that which is science's: a proven method for investigating our universe. But render unto the Bible the search for purpose and the poetry that describes the purpose.

With every new discovery, science produces a picture of a more wondrous world. As it does, it appeals to our sense of humility. If believers can get beyond the rigidities of their worst days—if the sorts of unjustified reactions against Galileo and Darwin can be finally put to rest—then perhaps secular scientists can open their minds to the possibility of purpose. With depth of understanding in these two sources of knowledge, we can discern that they complement each other rather than compete.

CHAPTER 3

The Age

of

Our Universe:

Six Days

and

Fifteen Billion Years

Discoveries of fossils from hundreds of millions of years ago make the news almost daily. And what of the multibillion-year age of the universe? I can't pick up a copy of the latest scientific journal without reading new evidence indicating that the heavens and all they contain appeared with a big bang some ten to twenty billion years ago. Yet the Bible hasn't changed its story. Six days is all we have to get from that beginning, the big bang, to humanity.

Of course God may have put the fossils in the ground to make the universe look old. While this answer may not satisfy all opinions, there is no way to disprove the thesis. And it may be true. But there is another approach to creation, one that retains a traditional view of Genesis while incorporating the discoveries of modern science. I share that answer with you here.

MEASURING THE AGE OF THE UNIVERSE

Scientific estimates of fossil ages and a multibillion-year-old universe come from a variety of diverse and independent measurements. Since the 1940s, carbon-14 has been a standard method for dating fairly young fossils, those with ages up to about thirty thousand years. But with C-14 there has always been room for suspicion. For the ages to be valid, the amount of radioactive carbon (i.e., C-14) in the atmosphere must always have been constant. Carbon-14 is produced by cosmic rays smashing into nitrogen atoms near the top of Earth's atmosphere. If the intensity of cosmic radiation was different in bygone ages, then the C-14 dating system will be in error. There is now evidence that cosmic radiation has not been absolutely constant during the eras that C-14 is used for fossil dating.

But with improvements in the sensitivity of instruments such as isotope separation mass spectrometers, fossil ages from tens of thousands of years before the present can be measured using the totally different radioactive decay series of uranium-thorium. This decay scheme is independent of cosmic radiation. The uranium-thorium method gives the same answer as does C-14. Well, not exactly the same—the calculated ages often differ by 10 or 20 percent. But when measuring the age of a thirty-thousand year-old bone, does it really make a difference if the age is not actually thirty thousand years but "only" twenty-four thousand years? That's still a long way from the biblical account of six days. For the dating of older fossils, such as those from dinosaurs or from the Burgess Shale, there are equally convincing, multiple independent methods.

The age of the universe can also be estimated using several unrelated methods, all of which yield data showing it to be between ten and twenty billion years old—again, a value not very similar to the six days of Genesis.

(The range in the estimated age of the universe arises from uncertainty about its past and present rate of expansion, and from estimates of the ages of individual stars.[1,2] Both sets of data rely on extrapolation of data far beyond the regions of measurement. For the purpose of this chapter, I will use the value of fifteen billion years.)

There is a simple answer to the problem of a scientifically old and biblically young universe, an answer that has within it the core of a complex truth. Time as described in the Bible may not be the same as we know time today. We find a hint for this in the 2,900-year-old Book of Psalms: "A thousand years in Your sight are as a day that passes, as a watch in the night" (Ps. 90:4). Perhaps from a biblical perspective the six days of Genesis include the fifteen billion years we earthbound mortals estimate to be the span of time since the beginning of time, just as a watch in the night might include a thousand years.

That makes sense . . . if you believe Psalm 90!

Skeptics should not be too eager to chortle over my use of Psalms as if I were seeking a rationalization for squeezing fifteen billion years into six days. I don't want a trivial correlation between science and theology such as "Let's call each day a very long epoch, lasting millions or even billions of years." The approach that "the six days were really six epochs" has scant biblical basis. Ancient commentaries, those written millennia before the discoveries of paleontology and cosmology disclosed any hints that the universe was billions of years old, state definitively that the six days of Genesis were twenty-four hours each, the total duration of which was "as the six days of our work week."[3,4]

But these same commentators continued and described those six twenty-four-hour days as containing "all the secrets and ages of the universe."[5] Now that sounds like Psalm 90:4!

Modern students of the Bible might prefer that the days of Genesis be epochs. That would accommodate the findings of cosmology and paleontology to a cursory reading of Genesis. But it smacks of apologetics. It may come as a surprise; however, the discoveries of science have made apologies for this seeming discrepancy between science and the Book of Genesis unnecessary. Deep within Psalm 90, there is the truth of a physical reality: the six days of Genesis actually did contain the billions of years of the cosmos even while the days remained twenty-four-hour days.

To grasp the full meaning of Psalm 90, we need a deeper understanding of the universe than casual observation reveals. The twelfth-century commentator Moses Maimonides advised that we study astronomy and physics if we desire to comprehend the world and God's management of it.[6] A superficial understanding of the former leads to misconceptions in the latter.

A superficial reading of the universe often fails to provide the truth. We see this every day in sunrise-sunset. Our simple perception of the Earth-Sun system is that Earth is at rest relative to the Sun and the Sun is circling Earth. Today, few believe this is the way the solar system is put together. The discoveries of astronomy have demonstrated that it is not the motion of the Sun that produces a sunrise, but instead, contrary to our every perception, the sensation of sunrise and sunset results from the Earth rotating from west to east on its axis once each twenty-four hours.

Now the implications of accepting this as true are quite amazing. If we are to get a day and a night out of twenty-four hours of rotation, the Earth must make a complete rotation once in each twenty-four-hour period. Earth is some 24,000 miles in circumference at the equator. To get those 24,000 miles all the way around in twenty-four hours, every point and every person on the equator must be moving at 1,000 miles per hour!

Of course you are probably reading this book quite a distance north or south of the equatorial tropics, so you can relax—a bit. At the latitude of, say, Atlanta or Adelaide or Israel (about 30° latitude), the Earth's circumference is only 21,000 miles. Your daily trip is a bit more leisurely: a mere 875 mph. And our day-night rotation is the small part of our cosmic travels. To get through a year in 365 days, the Earth moves around the Sun at 20 miles per second. And the entire solar system, us included, is hurtling around the center of our galaxy, the Milky Way, at ten times that speed.

Can you feel any of this? Are the clouds flying by at breakneck speed? No. So why believe it? Well, just as you can't believe everything you read, you can't believe everything you see. It takes research, intellectual effort, to find the truth. Our senses may be adequate for getting us to work and back, but when it comes to questions of the cosmos, our senses need help. The secrets of nature are not always revealed by a literal reading of nature.

What is true for the cosmos is also true for the Bible.

In the Book of Proverbs 2,800 years ago, King Solomon wrote: "A word well spoken is like apples of gold in bowls of silver" (Prov. 25:11). Maimonides, eight hundred years ago, developed the theme. Seen from a distance, only the silver bowl and its beauty are noticed. With closer inspection the more valuable apples of gold are discovered within. What is the bowl of silver about which Proverbs speaks?

It is the literal text of the Bible. Even with a superficial reading, seen from a distance, as it were, it is beautiful. The stories have delighted child and adult alike for thousands of years. Only close inspection, deep study of the text, reveals the golden apples, the subtleties, held within. As gold is more valuable than silver, Maimonides continued, so these subtleties, the quiet truths, expand the meaning far beyond a literal reading.

Let's look at a few of these golden apples and see if they can help us fit fifteen billion years into six twenty-four-hour days.

THE FIRST GOLDEN APPLE:
THE SIX DAYS THAT PRECEDED ADAM AND EVE

In the fall of the secular year 1996, the biblical calendar reached the year 5757. This is calculated by adding the ages of the generations of humankind as they are listed in the Bible and the rulers thereafter.

The calendar has reached 5757. But what event marks its beginning? Logically, the calendar should start with the creation of the world. That would be the generations since Adam plus the six preceding days. But such is not the case. Two thousand years ago, long before there was any controversy over hundred-million-year-old dinosaur bones and cosmic ages reaching into the billions of years, the starting date of the biblical calendar was set at the creation of the souls of humankind (Gen. 1:27), and not at the creation of the universe, the "In the beginning" of Genesis 1:1.[7-9]

To understand the basis upon which ancient scholars relied in excluding the six days of Genesis from the biblical calendar, I suggest reading the opening chapter of Genesis a few times, paying particular attention to the description of the events and the flow of time related to those events. Then read any other chapter in the entire Bible, again concentrating on the flow of events and the related flow of time. Note how the context changes. The description of time in the Bible is divided into two categories: the first six days and all the time thereafter.

During those six days, blocks of events are described and then we are told that a day passed. This is repeated in a totally objective fashion six times. It is as if there were a consciousness reporting the events from the outside looking in. There is no intimate relation between the events and the passage of time. The text, for example, does not state that five hours and twenty-five minutes into the third day God separated the water from the dry land; and then, after another nine hours and forty-five minutes, plant life appeared. Rather, we are told that the land and waters separated, plant life appeared, "And there was evening and there was morning a third day" (Gen. 1:9–13). No hint is given for the time each of these major events took.

With the appearance of humankind, the accounting changes dramatically. The events now become the cause of the flow of time. Adam and Eve live 130 years and are the parents of Seth (Genesis 4:25; 5:3). Seth lives 105 years and is father to Enosh (Gen. 5: 6). The passage of time is totally tied to the earthly events being described. These are indeed years of an earthly calendar.

Now here's a puzzle. If, as those ancient commentators claimed, the six days of Genesis are twenty-four-hour days, then why not include them in the calendar? Why not have the calendar start six days earlier? And why must these commentators tell me the days are twenty-four hours each? The Bible says "day." I know a day takes twenty-four hours to pass. Why did they think I would think otherwise?

As we learned, our questions were anticipated thousands of years ago.[10] The six days are not included in the calendar because within those (six twenty-four-hour) days are all the secrets and ages of the universe.

The confusion mounts. How can six days contain the ages of the universe? And if they are truly ages, then why refer to them as days?

The ancient realization that somehow the days of Genesis contained the generations of the cosmos is based on two biblical verses: "These are the *generations* of the heavens and the earth when they were created in the *day* that the Eternal God made earth and heavens" (Gen. 2:4); and "This is the book of the *generations* of Adam in the *day* that God created Adam" (Gen. 5:1). In both verses, generations are juxtaposed to days of Genesis.

If the six twenty-four-hour days of Genesis were adequate to include all the days of the universe, the cosmic flow from the creation at the big bang to the creation of humankind, we clearly require an understanding of time that is not obvious to our unaided senses. Albert Einstein provided that understanding.

THE SECOND GOLDEN APPLE: UNDERSTANDING TIME

In 1915, Einstein published a description of nature which revealed an extraordinary, and seemingly quite unnatural fact: the rate at which time passes is not the same at all places. Changes in gravity and changes in the velocity at which we travel actually change the rate at which our time flows. At first such a concept appeared to be highly speculative and so this aspect of nature was referred to as the *theory* of relativity. But it is no longer a theory. During the past few decades, the relativity of time has been tested and verified thousands of times. It is now the *law* of relativity. Einstein had discovered a hitherto overlooked law of nature.

If anything in our life seems constant, it is the flow of time. This perception, or rather this misperception, of time results from the reality that the events with which we are familiar all occur on Earth, or if not exactly on Earth, then quite close to Earth. Huge changes in gravity (G) or velocity (V) are required to produce easily measurable changes in the flow of time. And even with the needed large variations in G or V, the flow of time wherever you happen to be will always appear as normal, just as it does right now. It appears normal because you and your biology are in tune with the local system. Only if we view events across a boundary, looking from one location into another location that has a very different G or V, can we observe the effect of this extraordinary law of nature which was discovered by Einstein. The relativity of time is encountered only when comparing one system relative to another; hence the name the law of *relativity*.

The law of relativity tells us that the flow of time at a location with high gravity or high velocity is actually slower than at another location with lower gravity or lower velocity. This means that the duration between ticks of a clock (and the beats of a heart, and even the time to ripen oranges) in the high-G (or high-V) environment is actually longer than the duration between ticks of a clock (or beats of a heart) in the low-G (or low-V) environment. These differences in time's passage are known as time dilation.

If, at this point, you find yourself saying there is no way of com-
prehending how or why gravity can make time slow down, do not
despair. You are in the company of the other approximately five billion human inhabitants of this Earth.

We have no problem conceptualizing the effect of gravity on
weight. I weigh myself in the morning: 150 pounds on the bathroom
scale. Toting the same scale, I board a rocketship and zoom off to the
moon. Upon arrival on the moon, out comes my scale. I now weigh
a mere 25 pounds! Have I discovered the diet program to end all
diets, or been busy doing aerobics on the way up? Not likely. My
waistline is just what it was when I left Earth. So why my loss in
weight? Because, as we all know, the force of gravity is lower on the
Moon than on Earth. I am the same size as I was on the Earth. I
have the same *mass* as I had on Earth, but I now weigh only one-
sixth my Earth weight because the force of gravity on the Moon is
one-sixth that of the Earth's. And changes in the force of gravity
produce parallel changes in the *weight* of mass.

We sense this effect of gravity on weight each time we feel the
weightless sensation during the plunge of a roller coaster or stand in
an elevator and feel the floor push upward on the bottoms of our
feet as it starts its ascent. The sensation of increased weight on our
feet would be exactly the same if the elevator remained stationary
and instead, by some miracle, the force of gravity suddenly increased.
Subconsciously, our brain is recording this information.

Just as gravity affects the weight of mass, it also affects the flow of
time, but at a much less dramatic rate. That is why it took an Einstein
to discover this law of nature.

A few years ago at dinner, my family and I discussed the concept
of time dilation. We decided to take a mental excursion into the
realm of relativistic time to make the effect more clear. We can know
how it works even if we can't understand *why* it works.

We conjured an imaginary planet so massive that its gravity
slowed time by a factor of 350,000 relative to Earth's rate of time.
That meant that while we here on Earth live out two years, a mere
three minutes would tick by on that imaginary planet. My then
eleven-year-old daughter, Hadas, exclaimed, "Dad, this is great! This
is just super. Send me to that planet. I'll stay there for three minutes,

do two years of homework, come home and no more homework for two years!"

That's not quite correct, is it?

In Hadas' time, three minutes will have passed. But for us on Earth, those three minutes will have taken two years. In those two years Hadas will have done only three minutes of homework and aged only three minutes. Upon return to Earth, all her friends will be thirteen while she will still be eleven. That is the proven nature of time in our awesome universe.

Had I watched Hadas from my low-gravity location, her time (and all her events, including her aging) would have passed v e r y s l o w l y. To me, events in my system were totally normal. From Hadas' perspective, her watch and her actions were normal, but looking across the reaches of space from her high-gravity system into my lower-gravity system, she would have seen my watch and everything else on Earth going very rapidly. Between two beats of her heart, my heart would beat 350,000 times.

There was only one sequence of events, one Hadas. For her, that sequence took three minutes. Three minutes of heartbeats, three minutes of study and homework. For us here on Earth, that identical time span took two years. Two years of heartbeats, two years of life's accomplishments and joys, two years of orange harvests. And both occurred in exactly the same "time." Hadas lived three minutes while we here on Earth lived two years. And which time is correct? Both. It's all relative.

Let's apply this understanding of time to the age of the universe. The implications are profound and the conclusions surprising.

THE THIRD GOLDEN APPLE: THE MANY AGES OF OUR UNIVERSE

Each planet, each star, each location within our universe has its own unique gravitational potential, its own relative velocity and, therefore, its own unique rate at which the local proper time passes, its own age. If we were Moon people where the gravity is lower than on Earth, our clocks would tick a bit faster than the identical clocks on Earth. If we were Sun people where the surface gravity is thirty times greater than on Earth, and if we and our watches could stand the heat, we would observe that in one Earth year, the Sun-based clocks ticked off one year minus sixty-seven seconds.

This difference has actually been measured, but not with clocks sent to the Sun. There is no need for that. The Sun sends us its timing information in every beam of sunlight. The timepiece of the universe is not manufactured by a watchmaker, skilled though that craftsperson may be. The clock of the universe is the light of the universe. Each wave of light is a tick of the cosmic clock. The frequencies of light waves are the timepieces of the universe.[11-14]

Waves of sunlight reaching Earth are stretched longer by 2.12 parts in a million relative to similar light waves generated on Earth. That stretching of the light waves means that the rate at which they reach us is lowered by 2.12 parts per million. This lowering of the light wave frequency is the measure of the slowing of time. For every million Earth seconds, the Sun's clock would "lose" 2.12 seconds relative to our clocks here on Earth. The 2.12 parts per million equals 67 seconds per year, exactly the amount predicted by the laws of relativity.

There are any number of ages for our universe, each being correct for the location at which the measurement is made. And there are literally billions of locations where a clock, if we could place one there, would tick so slowly that fifteen billion Earth years would pass while it recorded only six twenty-four-hour days. So finding an equality between the six days of Genesis and fifteen billion Earth years is not a problem. But unfortunately, this simple solution is inadequate. Individual locations are not relevant to the opening chapter of Genesis. If we are to discover the basis behind the exclusion from the biblical calendar of those six days of Genesis, we must identify the universal perspective of the Bible's space-time reference frame for those eductive, six pre-Adam days.

THE FOURTH GOLDEN APPLE: A UNIVERSAL CLOCK

We know three facts *with complete certainty* about the description of time in the Bible:

1. The biblical calendar is divided into two sections: the first six days of Genesis and all the time thereafter. Those six days are not, and

never have been, included in the calendar of the years which follow Adam.

2. Time in the biblical calendar after Adam *must have been* Earth-based. Archaeology proves this. The radioactive dates of archaeological discoveries related to the post-Adam period, such as the early Bronze Age, the beginning of writing, the battle of Jericho, closely match the dates derived from the biblical calendar for those same events.[15] That radioactive decay occurred here on Earth in Earth time. Since the dates are a good match, the corresponding dates of the Bible must also use an earth-based calendar. There are no effects of biblical relativistic time dilation after Adam.

3. Most important of all, we know that there is no possible way for those first six days to have had an Earth-based perspective simply because for the first two of those six days there was no Earth. As Genesis 1:2 states "And the earth was unformed. . . ."

The *only* perspective available for the *entire* Six Day period is that of the total universe, one that encompasses the entire creation.★

What would happen to the flow of time if, instead of our seeking one specific location within the universe, we considered the universe in its entirety?

Fortunately, for our measure of times within the vast reaches of the cosmos, there is no need for space travel. Just as with time on the surface of the Sun, the cosmos sends us the needed information. It arrives neatly packaged in the form of wavelengths of light.

The following is a reader-friendly description of a cosmology of time. Subjectivity is difficult, perhaps impossible, to avoid. I have an agenda, to demonstrate a harmony between science and the Bible. To limit my subjectivity, I have restricted my sources of scientific information to peer-reviewed data that are accepted in physics laboratories of leading universities. I have limited my sources of biblical

---

★Occasionally I hear the complaint: "Why didn't the biblical calendar become earth-based on day three, when earth appeared?" This is a reasonable, though not a valid, question. The biblical calendar is a given and has been fixed for well over two thousand years. The biblical description of time is the same for the entire six-day period. It changes only at the end of the sixth day. The author of the Bible chose to use universal time, not earth-based time, until Adam. We are not here to rewrite the Bible. We are trying to understand it as it is.

interpretation to the Talmud (redacted in the year 500) and the kabalist Nahmanides (1250), the two mainstream traditional paths to the deeper meanings held within the text of Genesis. These sources were recorded centuries, even millennia, before the discoveries of modern science and so were not influenced by those discoveries. I have referenced the sources at each key juncture of the discussion. I will warn the reader when the modern theology begins. Until then it is pure, peer-reviewed physics and traditional Genesis.

LIGHT AS THE COSMIC CLOCK

Light has the mysterious property of being both a particle and a wave. It is the wave aspect that allows us to measure time over cosmic distances. What we refer to as visible light is only one particular band of wavelengths in a nearly infinite range of electromagnetic radiations all of which travel at the same speed: the speed of light (c = 300 million meters per second in a vacuum). The wavelength of the radiation determines whether it falls within our range of vision. We see radiation wavelengths of 0.0001 centimeters as the red end of the spectrum and 0.00001 centimeter wavelengths as the blue end. A microwave oven has radiation waves that are approximately 10 centimeters long, while gamma rays are shorter than 0.000000001 centimeters. The shorter the wavelength, the higher the wave frequency and the higher the wave energy. The velocity remains constant.

Our application of light in calculating the passage of time on the Sun relative to the Earth demonstrated the usefulness of light frequency as a cosmic clock.[16] It related time at one location in the universe (the Sun) to time at another location (the Earth).

A common error in exploring the brief biblical age of the universe relative to the discoveries of cosmology is to view the universe from a specific location rather than choosing a reference frame that embraces the entire universe and retains that universal perspective for the entire six days. The clock of Genesis starts with the creation of the universe and continues till the creation of humankind. It must identify the relative passage of time not between particular *places* in the universe but between *moments* in the universe as the universe evolved from the big bang.

Just after the big bang, the universe was a concentrated hot plasma with nearly identical energies throughout. The relative passage of time varied only slightly, if at all, among its components. But as the universe expanded and cooled, vastly different local gravities and velocities evolved, having vastly different rates at which local proper times flowed. For our understanding of Genesis time, we must maintain the undifferentiated frame of reference that pervaded the universe at its beginning.

The lights we see in the heavens originate with energy released in stellar and galactic nuclear reactions. But there is another source of cosmic radiation, one that has been present since the creation of the universe. That is the radiation remnant, the echo as it were, of the big bang. This cosmic background radiation (CBR) fills the entire universe, unrelated to any particular source. Discovered by Arno Penzias and Robert Wilson in 1965, it is the only source of radiation that has been present and ubiquitous since the creation. CBR frequency forms the basis of cosmic proper time, the biblical clock of Genesis.

Concepts of cosmic proper time relative to the expansion of the universe and its perceived age have been presented in such prestigious peer-reviewed journals as *Nature* and the *American Journal of Physics*. Cosmic proper time does not replace conventional time but rather augments it.[17-21]

Let's apply this cosmic timer to the universe.

THE FIFTH GOLDEN APPLE: WHEN THE UNIVERSE WAS SMALL

Three aspects of the universe produce identical effects on radiation frequency. Positive differences in velocity, gravity, and the stretching of space as the universe expands all increase (stretch) the wavelength of radiation. Since the frequency of radiation (and hence the beat of the cosmic clock) is lowered in direct proportion to the increase in wavelength, this increase in wavelength slows the perceived passage of time. The first two of these three phenomena relate to differences in the flow of conventional time—biological time—between specific locations. The third, the universal stretching of space, equally alters the perception of time's flow as reckoned by the universal cosmic clock.

It is standard practice to label an epoch of the universe by the expansion factor [the ratio of stretched radiation wavelength as it is observed relative to the unstretched wavelength as it was emitted. This is] defined by redshift, z, even when an epoch is so early the redshift cannot be observed in detected radiation. . . . The standard interpretation of the redshift as an effect of expansion of the universe predicts that the same redshift factor applies to observed rates of occurrence of distant events.[22]

At this stage, kabalah and cosmology have merged. If we can understand how the expansion of the universe since the big bang has altered the frequency of cosmic background radiation, we may be able to understand how the six days of Genesis "contain all the secrets and all the ages of the universe." In essence we seek to map cosmic/Genesis time onto time as we perceive it in our corner of the universe.

At the big bang, our entire visible universe was packed into a minuscule speck of space. Since then, expansion of the universe and the inherent stretching of space has moved the "edge" of the universe out by billions of light years. Considering that each light year, the distance light travels in one year, is ten million million kilometers, it is clear that we live in a huge universe. The frigid (-270°C) cosmic background radiation observed in all directions of the sky is the stretched radiation left from the immense heat of the big bang when the universe was tiny.[23-25]

This immense stretching of space since the big bang has strong implications for our cosmic clock. Waves of radiation that have propagated in space since the early universe have been stretched, expanded, by the same proportion that the universe has expanded.[26] For example, as the universe doubled in size, the distance between wave crests (and hence the time between ticks of its clock) also doubled as the wave was stretched by the expanding space. (see Figure 2). For that clock, time would be passing at half its original rate.

The primordial explosion by which the universe was created, the big bang, did not produce matter directly. It produced a pure, exquisitely hot radiant energy of such high level that matter was able to form from the energy. This was the transition of E (energy) into m

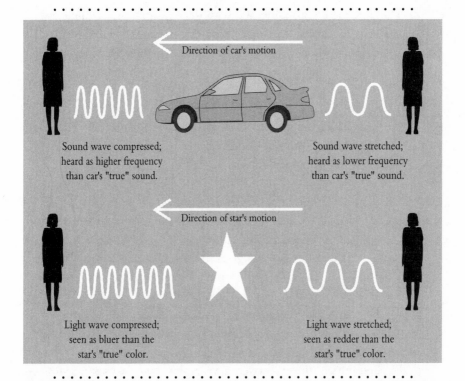

Direction of car's motion

Sound wave compressed; heard as higher frequency than car's "true" sound.

Sound wave stretched; heard as lower frequency than car's "true" sound.

Direction of star's motion

Light wave compressed; seen as bluer than the star's "true" color.

Light wave stretched; seen as redder than the star's "true" color.

FIGURE 2: The effect of relative motion and the stretching of space on the observed frequency of light waves

In our relativistic universe, the condition of the observer relative to the events being observed strongly affects the perception of those events! The expansion or contraction of space produces the identical effect on the frequency of a light wave as does relative motion between light source and observer. As the space of the universe expands, light waves are stretched *(right side of figure)*. If the space of the universe contracts, then a light wave traveling through space will be similarly contracted *(left side of figure)*.

Starlight arriving from galaxies in deep space is stretched, with the extent of stretching increasing as a function of the galaxies' increasing distance from us. The implication is that our universe is expanding and has been expanding for some 15 billion years.

(matter) of Einstein's famous $E = mc^2.$ The c^2, the speed of light squared or multiplied by itself, is a huge number and implies that even a tiny amount of matter contains a huge amount of energy and therefore requires a huge amount of energy to form.

The universe is expanding, and with the expansion it is cooling as the initially concentrated energy becomes ever more "dilute" within the ever larger volume. Soon after the big bang, the energy level, as indicated by the ambient temperature, fell below the minimum value that allowed energy to change into matter as we know it. This transition value, the threshold temperature of protons and neutrons, marks the moment of what I refer to herein as *quark confinement*. After quark confinement, protons and neutrons—the building blocks of matter as we know it—could no longer form.

Nahmanides, in the year 1250, described the process with uncanny accuracy: the initial creation produced an entity so thin it had no substance to it. It was the only physical creation ever to occur and was all concentrated within the speck of space that was the entire universe just following its creation. (This seven-hundred-year-old insight could be a quote from a modern physics textbook.) As the universe expanded from the size of that initial minuscule space, the primordial substanceless substance changed into matter as we know it. Biblical time, he continued, starts ("grabs hold," in his words) with the appearance of matter.[27]

Physics, during the past twenty years, has come to agree with Nahmanides. Nahmanides writes that his teachers learned this account from the first word of the Bible, *beraesheet,* which means "In the beginning of." In the beginning of what, they asked? In the beginning of time was their conclusion.

Nahmanides' insight that biblical time takes hold with the appearance of matter is quite extraordinary. Visible light rays, invisible microwaves, X rays, gamma rays are all forms of the same type of radiant energy known as electromagnetic radiation. As science has discovered, radiant energy does not experience the flow of time. Radiant energy, such as the light rays you are seeing this very moment, exists in a state that might be described as an "eternal now," a state in which time does not pass. (This is a concept we can write but not intellectually grasp because all our existence is within the flow of time.) Time, in the sense that we experience it, is totally related to the material world. Time truly takes hold when matter forms.

That transition from energy to stable matter occurred 0.00001 seconds after the big bang.[28] The universe was then approximately a million million times smaller and hotter than it is today.[29]

THE SIXTH GOLDEN APPLE: THE AGES OF OUR UNIVERSE

"And the earth was *tohu* and *bohu* . . ." (Gen. 1:2). The usual translation of this verse from the book of Genesis is "And the earth was unformed (*tohu*) and void (*bohu*)." Unformed or chaotic is a fair translation of *tohu*. But *bohu* does not only mean void. Both the Talmud and Nahmanides state that *bohu* means filled with the building blocks of matter.[30,31]

A more accurate, though cumbersome, translation of Genesis 1:2 is: "And the earth was in a state of chaos but filled with the building blocks of matter."

Since biblical time takes hold with the appearance of matter, the biblical clock starts at *bohu*, that instant just after the big bang when stable matter as we know it formed from energy. The age of all matter in the universe dates back to *bohu*, the moment of quark confinement.

We know the temperature and hence the frequency of radiation energy in the universe at quark confinement. It is not a value extrapolated or estimated from conditions in the distant past or far out in space. It is measured right here on Earth in the most advanced physics laboratories and corresponds to a temperature approximately a million million times hotter than the current 3°K black of space. That radiant energy had a frequency a million million times greater than the radiation of today's cosmic background radiation.

The radiation from that moment of quark confinement has been stretched a million-millionfold. Its redshift, z, as observed today is 10^{12}. That stretching of the light waves has slowed the frequency of the cosmic clock—expanded the perceived time between ticks of that clock—by a million million. "This also applies to proper rates of events as one sees by the application of a sequence of Lorentz time-dilation factors."[32] Those are solid values in physics.[33–36]

To measure the age of the universe, we look *back* in time. From our perspective using Earth-based clocks running at a rate determined by the conditions of today's Earth, we measure a fifteen-billion-year age. And that is correct for our local view. The Bible adopts this Earthly perspective, but only for times after Adam. The Bible's clock before Adam is not a clock tied to any one location. It is a clock that looks *forward* in time from the creation, encompassing the entire universe, a universal clock tuned to the cosmic radiation at the moment when matter formed. That cosmic time-piece, as observed today, ticks a million million times more slowly than at its inception. The million millionfold stretching of radiation since *bohu* caused that million-million-to-one ratio in this perception of time.

This cosmic clock records the passage of one minute while we on Earth experience a million million minutes. The dinosaurs ruled the Earth for 120 million years, *as measured by our perception of time.* Those clocks are set by the decay of radioactive nuclides here on Earth and they are correct for our earthly system. But to know the cosmic time we must divide earth time by a million million. At this million-million-to-one ratio those 120 million Earth years lasted a mere hour.

That's the peer-reviewed physics and the biblical tradition of this discussion. Now for the modern theology.

What does all this mean for the age of the universe?

In terms of days and years and millennia, this stretching of the cosmic perception of time by a factor of a million million, the division of fifteen billion years by a million million reduces those fifteen billion years to six days!

If the universe had been any other size, temperature, or mass, or the threshold temperature of matter (protons and neutrons) had been different, this relationship would not exist. Cosmologists are in awe that the mass and the energy of expansion of the universe are matched with the "incredible fine-tuning" of one part in $10^{120.}$ It is almost as if the values had been selected. Perhaps they have.

Genesis and science are *both* correct. When one asks if six days or fifteen billion years passed before the appearance of humankind, the correct answer is "yes."

The correspondence between biblical time and the cosmic clock might be seen as fortuitous. There is, however, a test to this theory. The Bible not only tells us that six days passed between the creation of the universe and the creation of Adam, but it also records the key events that occurred on each of those days. With this in mind, let's compare day by day the fidelity by which the events of Genesis map onto the corresponding discoveries of science. I think even an avowed skeptic, secular or creationist, will find the match too good to be relegated solely to fortuity. (For a simplified analogy of cosmic background radiation, CBR, as a universal clock, please see the appendix, section a.)

. . . .

CHAPTER 4

The

Six Days

of

Genesis

The Duration of the Six Days of Genesis		Blueshift (z+1)	
From the Bible's perspective looking forward in time from start of day one	From Earth's perspective looking backward in time from the present	From Bible's perspective at the start of day one	Approximate years before Adam at start of each day
Day one 24 hrs	8 billion yrs	1	15 3/4 billion yrs
Day two 24 hrs	4 billion yrs	2.0×10^{12}	7 3/4 billion yrs
Day three 24 hrs	2 billion yrs	3.0×10^{12}	3 3/4 billion yrs
Day four 24 hrs	1 billion yrs	3.5×10^{12}	1 3/4 billion yrs
Day five 24 hrs	1/2 billion yrs	3.7×10^{12}	3/4 billion yrs
Day six 24 hrs	1/4 billion yrs	3.9×10^{12}	1/4 billion yrs
Near end of day six		4.0×10^{12}	
Total: Six 24-hour days	15-3/4 billion years		

The general correlation between Genesis cosmic time and Earth-based time, presented in the preceding chapter, is surprisingly good. The day-by-day details of that temporal harmony are intriguing. In deriving those details, peer-reviewed science and traditional biblical commentary alone are my sources. I urge the scientifically disinclined reader to suffer through or skip over the bit of mathematics that follows and continue on to the day-by-day comparison of science with Genesis 1.

The million-million-factor difference between our local perception of time and Genesis cosmic time is an average for the six days of creation. As discussed, it derives from the approximate million-millionfold stretching of light waves as the universe expanded.[1-3] My use of the term light wave includes all energies of electromagnetic radiation (microwaves, X rays, gamma rays, etc.), and not merely the narrow range of wave frequencies and wavelengths visible as light.*

The universe is filled in all directions with radiation that is, in essence, the echo of the big bang. It forms a background "noise" unrelated to any particular location and is the only radiation that has been present since the big bang. Termed the cosmic background radiation (CBR), it is uniquely characterized as the temperature of a black body that would emit the same radiation pattern. The higher the temperature of the body, the higher the frequency of the radiation. For simplicity, it is customary to refer to the CBR frequencies that fill the universe as the corresponding black body temperature of the universe. That temperature today is 2.73°K (approximately minus 270°C). It is, in a sense, the temperature of the black of space.

CBR is the clock of the cosmos.[4-7] Its wave frequency is the rate which the cosmic clock ticks. "The directly measurable coordinate along the line of sight [into space] is not time, but redshift *(z)*"—the ratio of CBR frequencies at some distance in the past relative to CBR frequencies observed today.[7] Just after the big bang, when the universe was vastly more compact, all the radiation spread throughout today's huge universe was pressed within a small primordial space. This immense concentration of energy resulted in CBR tem-

---

*The exceptions to the direct relationship between the stretching of space and the stretching (lengthening) of light waves were during phase transitions or matter-antimatter annihilation, such as electron-positron annihilation. The energy released by these events momentarily raised the temperature of the universe. Higher temperature corresponds to higher radiation energy, which in turn relates to shorter wavelength and higher wave frequency for the radiant energy.

peratures and wave frequencies million upon million times greater than that of the frigid 2.73°K of space today. The cosmic clock then "ticked" much more rapidly than it does today.

As the universe expanded, its size (scale) and temperature, and therefore its clock, were becoming ever more similar to that of our current universe. Because of this, the "duration" of each successive twenty-four-hour Genesis day encompassed a span of time ever more similar to time as reckoned from our Earth-based perspective. Each doubling in size "slowed" the cosmic clock by a factor of 2. Since the time required for the universe to double in size increased exponentially as its size increased, the fractional rate of change in the cosmic clock (relative to Earth time) decreased exponentially. The task of this chapter is to quantify the exponential relationship between the twenty-four hours of each Genesis day and time as we perceive it on Earth.

The opening chapter of Genesis acts like the zoom lens of a camera. Day by day it focuses with increasing detail on less and less time and space. The first day of Genesis encompasses the entire universe. By the third day, only Earth is discussed. After day six, only that line of humanity leading to the patriarch Abraham is in biblical view. The Bible realizes the entire universe still exists. But its interest now rests solely on one line of humanity. This narrowing of perspective, in which each successive day presents in greater detail a smaller scope of time and space, finds a parallel in scientific notation.

When data range over many orders of magnitude, the information is often displayed on a logarithmic graph. Each factor of ten, be it from one to ten or from one million to ten million, is given equal space on the graph. Thus the first inch on the graph may represent 1,000 to 100, the second inch 100 to 10, the third inch 10 to 1. Following this example, the first inch covers 900 units (1000 minus 100), the second inch 90 units, the third only 9. If this approach were not used, the information contained in the last section would have to be squeezed into 9/900, or 1/100 of an inch, giving no clarity. Alternatively, the graph might extend for 100 inches (8 feet). Neither method is convenient. The solution is to present the data on a logarithmic scale.

Genesis has chosen this common mathematical technique to display in words the events of that vast span of time from the creation of the universe to the creation of the soul (not the body) of humankind. Each successive Genesis day exponentially represents fewer years as perceived from our earthly vantage. But the biblical relationship for time is not the artificial ten-based log system used in the above example. Genesis has chosen a base that occurs throughout the universe, a base known in mathematics as the natural log, *e*.

The graceful curve of the Nautilus seashell occurs in nature more often than any other shape.[8] Its lines trace out an exponential spiral. Each of the spiral's successive swirls is wider by a seemingly arbitrary but fixed factor. We see the curve repeated in the distribution of seeds on a sunflower, the curves of tusks, and the spread of stars in spiral galaxies, our own Milky Way included. Graphically, it also describes the relation between Genesis time and Earth time as the universe expanded out from its point-like creation of the big bang (see Figure 3).

Jakob Bernoulli, a contemporary of Isaac Newton, formulated the shape of this curve in the mathematical terms of polar coordinates:

$$r = e^{aH}$$

Here, *r* is the distance from the *x–y* intersection on a graph (the origin, *0*); *a* is a constant; *H* is the angle between the line *0r* and the *x*-axis. As *H* increases in value, the point *r* spirals out, ever further from the origin. When the exponent is negative, as in:

$$r = e^{-aH}$$

the point *r* spirals in toward the origin in a swirling vortex that has no end, always moving closer to, but never reaching, the origin. Yet the spiral is finite in length.

To the mathematically uninitiated, this discussion may seem complex. But bear in mind that Einstein's colleague, Sir James Jeans, marveling at how well the universe can be described mathematically, is quoted as having remarked "God is a mathematician."

As the value of pi (3.14159 . . .) is basic to descriptions of all circles (pi times the diameter of a circle equals its circumference), so *e* is basic to all spirals. In fact, a circle can be seen as a special case of a

Nautilus from Capetown, modern

Ammonite, Southern England,
270 million years

Sunfllower

Spiral Galaxy

FIGURE 3: The exponential spiral: a curve commonly found in our universe *(Nautilus and ammonite photos by Karen Benzion; sunflower, Hadas Schroeder; spiral galaxy, Hale Observatory)*

spiral where the rate of growth is zero. Appearing throughout nature and mathematics, e's value is the limit of

$$(1 + 1/n)^n$$

This is the well-known compound interest formula. For large values of n, it equals 2.71827. . . , a sequence that never ends, never repeats, and never exceeds 2.71828.

This exponential relationship may also be plotted on the more familiar linear *x–y* Cartesian coordinates as the formula:

$$A = A_0 e^{-Lt}$$

Since we are seeking the relationship between the *linear* flow of time as experienced on Earth and the cosmic time of Genesis, this Cartesian expression, rather than the polar equation, is relevant here. The identical equation describes the exponential rates of decay of radioactive atoms.

We measure the age of the universe to be some fifteen billion years old. That is our earthly perspective looking *back* from the present into the past. But the Bible's perspective is one that looks *forward* from the beginning. We learn this from the text's recording the first day as day one and not as the first day. Though the comparative, ordinal form of the numbers was used for all the other days (second, third, etc.), Genesis used the absolute cardinal form for day one because it was viewing time from the beginning of time, a perspective from which there was no other time for comparison.[9] As Nahmanides wrote, "when matter forms time grabs hold."[10]

That moment when stable matter as we know it formed, the moment of quark confinement, occurred 0.00001 second after the big bang. The 0.00001 second seems a small difference, but in that brief span, the universe transformed the pure energy of the creation into the building blocks of 100 billion galaxies, each containing 100 billion stars. Near at least one of those stars, a planet brims with life eager to know its origins.

In our application of the formula,

$$A = A_0 e^{-Lt}$$

A is Earth time, the number of current Earth days contained within one Genesis day at any given instant during the six-day period.

A_0 is the instantaneous ratio of cosmic time at the moment of quark confinement to current time or, equally stated, the ratio of cosmic background radiation frequency then to now, which in turn is proportional to the ratio of cosmic background temperature then[11] (10.9×10^{12}°K) to now[12] (2.73°K).

L is the natural log of 2 (usually written ln 2, and equals 0.693) divided by the half period.

The half period is one Genesis day.

t is time in Genesis days and increases from zero to six.

To learn the duration of each Genesis day as viewed from our perspective of time, we integrate the equation and solve this integrated form for each of the six days of Genesis:

$$\text{Integral of } A = (-A_0/L)e^{-Lt}$$

The units are days. Converting Earth days to Earth years (dividing *A* by 365) gives the data listed in the table at the start of this chapter.

The sum of the durations of all six days is a prediction of the age of the universe. This calculated age of the universe equals 15.75 billion years. It is a near duplicate of the estimated 16-billion-year age of the oldest stars.[13-15] Keep in mind that the scientific data for cosmic ages and temperatures are estimates and therefore not exact.

In addition to providing a calculated age of the universe, having a time period for each day permits us to compare the sequence of scientific observations of cosmology and paleontology with the events of each day as described in Genesis. The match, day by day, between the biblical description of our cosmic genesis and the description provided by science is extraordinary. It is summarized in the table on p. 67.

Genesis day one began 15.75 billion years ago and ended 7.75 billion years before present (B.P.). After two thousand years of denying the possibility, cosmology finally has come to inform us that during this period there was a beginning. Then as the temperature of the universe cooled, electrons were able to bond with atomic nuclei and light literally broke free. More time passed and galaxies started to form.

The Bible limited its description of this period to the creation, and then to light separating from darkness (Gen. 1:4). It mentions here an event that appears nowhere else in the entire Hebrew Bible, a one-time phenomenon described as the spirit of God hovering over the universe (Gen. 1:2). Biblically, this was required to start the making of the universe following its creation. Physics also calls for a one-time phenomenon which it named inflation. This was a sudden and brief expansion of the universe immediately following the big

The Six Days of Genesis

Day number	Start of day (years B.P.)	End of day (years B.P.)	Main event(s) of the day Bible's description	Scientific description
One	15,750,000,000	7,750,000,000	The creation of the universe; light separates from dark (Gen. 1:1–5)	The big bang marks the creation of the universe; light literally breaks free as electrons bond to atomic nuclei; galaxies start to form
Two	7,750,000,000	3,750,000,000	The heavenly firmament forms (Gen. 1:6–8)	Disk of Milky Way forms; Sun, a main sequence star, forms
Three	3,750,000,000	1,750,000,000	Oceans and dry land appear; the first life, plants, appear (Gen. 1: 9–13); kabalah states this marked only the start of plant life, which then developed during the following days	The earth has cooled and liquid water appears 3.8 billion years ago followed almost immediately by the first forms of life: bacteria and photosynthetic algae
Four	1,750,000,000	750,000,000	Sun, Moon, and stars become visible in heavens (Talmud Hagigah 12a) (Gen. 1:14–19)	Earth's atmosphere becomes transparent; photosynthesis produces oxygen-rich atmosphere
Five	750,000,000	250,000,000	First animal life swarms abundantly in waters; followed by reptiles and winged animals (Gen. 1: 20–23)	First multicellular animals; waters swarm with animal life having the basic body plans of all future animals; winged insects appear
Six	250,000,000	approx. 6,000	Land animals; mammals; humankind (Gen. 1:24–31)	Massive extinction destroys over 90% of life. Land is repopulated: hominids and then humans

bang. In a minuscule fraction of a second, the universe stretched from its point at creation to a size similar to that of today's solar system. Shortly thereafter, it is thought, the expansion "settled" into a rate similar to the currently measured value.

Day two began 7.75 billion years B.P. and ended 3.75 billion years B.P. During that period most of the stars of the Milky Way's spiral formed. The Sun, a main sequence star located in the spiral, formed 4.6 billion years ago. The oldest stars of the Milky Way are found in globular clusters, outside the spiral disk, and are barely visible to unaided vision.

The Bible tells that on day two the heavenly firmament took shape (Gen. 1:8). From our vantage, the heavens we see are almost totally composed of stars of the Milky Way's main spiral disk.

Day three started 3.75 billion years B.P. and ended 1.75 billion years B.P. From geophysical evidence of weathered rocks, we learn that Earth had cooled and liquid water appeared on it 3.8 billion years ago.[16] Contrary to scientific opinion held until recently, fossil data have demonstrated that the first simple plant life appeared immediately after liquid water and not billions of years later.[17]

The Biblical account of this period mentions the first liquid water on Earth as the oceans and dry land appear (Gen. 1:9). This is followed by plant life (Gen. 1:11). Though a simple reading of the text implies that all types of plants appeared on this day, kabalah corrects this misunderstanding, stating that "there was no special day assigned for this command for vegetation alone since it is not a unique work."[18] Of all the events listed for the six days, this is the only one that tradition states occurred over an extended period not limited to that particular day. A more detailed discussion of day three appears in the appendix.

Day four began 1.75 billion years B.P. and ended 750 million years B.P. The earth sciences have revealed data indicating that during this period the atmospheric concentration of photosynthetically produced oxygen rose to concentrations comparable with today's atmosphere.[19] There are indications that with the further cooling of Earth and the rise in atmospheric oxygen, the atmosphere, formerly translucent, became transparent.

The Bible relates that the Sun, Moon, and stars are either made during this period (Nahmanides) or that they were already present, but became visible during this period (Talmud).[20] With either scenario, the text tells us the heavenly bodies became so clearly visible (transparent atmosphere) that they could be used for telling times and seasons (Gen. 1:14–19).

Day five began 750 million years B.P. and ended 250 million years B.P. We have entered the Cambrian period. Paleontology now becomes the dominant science related to biblical commentary. The fossil record reveals the sudden, explosive appearance of the first animal life as it flourished in the oceans, 530 million years ago, simultaneously bringing into being all basic body plans of modern life.[21–23] Then approximately 360 million years ago in rapid succession amphibian reptiles and winged (insect) life appeared.

Here Genesis records the first mention of animal life. We are told that the waters *swarmed abundantly* with many types of animals. Then reptiles and winged animals appeared (Gen. 1:20, 21).

Day six began 250 million years B.P. and ended just under 6,000 years ago, at the time of Adam. Paleontology records that approximately 250 million years ago, there was a mass extinction of 90 percent of life followed by repopulation. Animal life then flourished on dry land, leading to mammals and culminating with hominids.

The Bible tells that animals populated the land, with mammals and finally humankind appearing. Thus ends the account of the six days, all of it packed into thirty-one verses.

The biblical choice for the timing of each Genesis day is in itself interesting. The opening of each day is heralded by a cosmic or global punctuation of major significance.

The start of day one, some sixteen billion years ago, marks the creation of the universe, the big bang. Day two opens at approximately eight billion years B.P., one of the dates estimated for the shaping of the galactic disk of the Milky Way. (The date is speculative and not all cosmologists agree on it.) The third day begins 3.8 billion years ago. This date betokened the close of an era during which Earth was bombarded by a rain of meteors so intense as to have made the start or survival of life highly improbable. Immediately, at that date, the first liquid water and the first traces of life appear.

At 1.8 billion years ago, the start of day four marked the beginning of eukaryotic life—life forms having cells with an inner nucleus containing most of the cell's genetic material (DNA). Prior to this time all life was prokaryotic—having cells without nuclei. All life forms larger than one-celled organisms such as bacteria are eukaryotic. Day five, starting 750 million years ago, timed the appearance of the first clearly multicellular organisms. Decimation, in the fullest meaning of the word, occurred at the start of day six, 250 million years B.P. Between 90 and 95 percent of all marine life disappeared from the fossil record at that date, setting the stage for the flourishing of animal life on dry land.

The Bible relates in thirty-one verses, in a few hundred words, events spanning sixteen billion years. These are events about which scientists have written literally millions of words. The entire development of animal life is summarized in eight biblical sentences. Considering the brevity of the biblical narrative, the match between the statements and timing in Genesis 1 and the discoveries of modern science is phenomenal, especially when we realize that all biblical interpretation used here was recorded centuries, even millennia, in the past and so was not in any way influenced by the discoveries of modern science. It is modern science that has come to match the biblical account of our genesis.

Scientists at times get annoyed with discussions like this one. It can be construed to imply the following syllogism: (a) Science confirms the Bible's wisdom. (b) The Bible already has all the answers we need relative to an understanding of humanity's place in the scheme of existence. Therefore, (c) science is unnecessary as a tool for understanding man's place in the scheme of existence. As a lifelong scientist, I would hardly want to encourage such a nonsensical line of deductive reasoning. Though the Bible is eerily true and filled with wisdom that would not have been known widely, if at all, when it was written, nowhere does it claim to have all the answers. The Bible may be the primary source for claiming that a purpose underlies our existence. But understanding the cause of that purpose can only be found, as Maimonides stated so many centuries ago, in a knowledge of the physical world. For that knowledge, the theologian must turn to the scientist.

Science and Bible are complementary, not mutually exclusive.

Knowing the laws of nature that strip the blue from a yellow sun and with it color the daytime sky or turn that same sun a brilliant orange-red at every dawn can only increase one's sense of awe at the intricacy of the creation. The discoveries of science have brought forth the *measurable* indication that our world has an aspect that exceeds the materialistic.

With the insights of Albert Einstein, we have discovered in the six days of Genesis the billions of years during which the universe developed. The Bible claims that the Creator remained active in the affairs of the creation. Before we search for that divine involvement, it behooves us to know the biblical concept of how this transcendent immanence is made manifest. Before we look for God in nature, we had better know the nature of God.

CHAPTER 5

The

Nature *of* God:
Biblical
Expectations
for an Infinite
yet Immanent Creator

The god an atheist does not believe in is usually not the God of the Bible. Unfortunately, the god of the "believer" is also often not the God of the Bible.

For millennia holy warriors have proclaimed priority rights to knowledge of God's supposed word. In Jericho and Ay, Mecca and Medina, Jerusalem and Tiberias, their conquests soaked the streets with what they considered to be secular blood. If we are seeking a rapprochement between the secular view of the world and that of the Bible, it would be instructive to know what the biblical God is supposed to be.

Shortly after the death of a colleague's mother, a visitor asked my bereaved friend if he could still believe in God. His reply was not what the guest expected: "As far as I know, Abraham died, Isaac died, Jacob died, Moses died, Noah died. Even Adam died. Every person in the Bible died. Every one who is born is destined to die. Nature and the Bible are in total agreement on that point."

That visitor is not alone in having constructed the bizarre concept that suffering somehow contradicts the biblical claim for a beneficent God. We can concoct a host of preconceived notions that describe an infinite Creator. Unfortunately, our notions are not necessarily relevant. If we want to know the traits of the biblical God, the place to look is the Bible. If we are investigating whether or not the Bible's description of God is realistic, the place to look is the world.

We are told that two aspects of a transcendent infinity labeled as God have been "condensed" and projected into our spatially and temporally finite universe. One is a revelation, written in words, that we call the Bible. The other reveals itself in the wonders of nature and the laws by which nature functions. That's all we know of this Creator. The world is only the silhouette or image (*tslm* in Hebrew) of the Creator (Gen. 1:27). The full essence of Infinity is beyond our ken (Ex. 33:23).

The Bible is primarily narrative, teaching through the lives of the biblical personages. From the recorded events, we can extract the Bible's concept of how the biblical God is made manifest in our world.

Immediately the Bible sets the basic format: nature was and will be the vehicle for most divine interactions within the world. Over the six days of Genesis (Gen. 1:1–31) the universe developed from its chaotic beginning, through the start of life, and on to the appearance of humankind. For this entire epoch, the word creation is used only for three events. The vast majority of events in that period were directed by the laws of nature, created as inherent parts of the universe. Surely an infinite Creator did not require six days or even six pico-seconds to produce the universe we know. Why not an instant universe? At the Israelite Exodus from Egypt, the Bible tells us that God used a wind that blew all night to split the Sea of Reeds (Ex. 14:21). Why an all-night wind rather than perhaps a hand from heaven? That would have been much more impressive.

The six days of Genesis and the wind at the sea ensure that the world will look natural. A natural world allows us to maintain our free will. The biblical Creator may be omnipotent, but in this universe each member of mankind chooses his or her own path. And sometimes we are given the opportunity to "persuade" even the Creator which path to choose.

The Israelites danced around a golden calf at the base of Mount

Sinai, the mountain upon which Moses, at that moment, was receiving the Ten Commandments. When God told Moses that their acts of idolatry warranted their destruction, Moses argued with God: if God destroyed Israel, the Egyptians would say God only took the Israelites out of Egypt to slay them in the desert. This would be counterproductive to the image of a beneficent Creator of the universe. God agreed and retracted the plan of national destruction (Ex. 32:14).

A few months later, the pardoned Israelites were about to enter the Promised Land. God told Moses to send scouts to reconnoiter the land (Numbers 13:11). The scouts returned with a report that the land was beautiful but its inhabitants were giants, far too powerful to defeat. The people believed the scouts. They wept and rebelled, seeking to return to Egypt. Again, a godly plan for national liberation. Again, Moses pointed out the bad publicity this would bring to the truth of ethical monotheism, and again God relented (Num. 14:20).

Is this a biblical hint that the Creator is limited? Of course not. We are, however, being informed that, according to the Bible, the universe was created with certain divinely chosen ground rules. Except for a very few exceptional cases, God has chosen to adhere to them.

The exceptions are seen as miracles, excursions from the norm. Bizarre as it sounds, miracles, once beyond the pale of an enlightened society, now have a scientific basis. Quantum mechanics has changed our understanding of nature. Not only are miracles theoretically possible according to QM, they are observed regularly in physics labs. In the pristine air of academia they are referred to as "insufficiently caused events," events that can be observed but that cannot be explained by the conditions that preceded them. That's the age-old biblical definition of a miracle.

Though miracles may be momentarily impressive, the Bible is well aware that they are not the path to impressing an awareness of the Divine. It took ten before Pharaoh believed something out of the ordinary was happening to his nation. And even then, just days after he had sent out the Israelites in the Exodus, he and his army chased after them (Ex. 14:6). The wonder of the miraculous lasts about twenty-four hours before it is rationalized as a chance act of nature. Nature's seemingly omnipotent independence challenged even the belief of those who, according to the Bible, spoke with God.

When God told ninety-nine-year-old Abraham that he and Sarah would conceive a child to carry on his lineage, Abraham doubted the possibility (Gen. 17:18). Instead, Abraham suggested to God that a more realistic approach would be for God to rely on Ishmael, the son of Sarah's maid. Sarah, then eighty-nine, laughed incredulously at the news (Gen. 18:12). God's reply was short and to the point: "Is anything too hard for the Eternal?" (Gen. 18:14).

The Bible tells us that "The Eternal spoke with Moses face to face as a person speaks to a friend" (Ex. 33:11). At God's command, Moses had announced the miraculous plagues in Egypt. Yet when God promised to supply meat for the Israelites as they marched through the desert, Moses' first reaction was not gratitude, but doubt. "And Moses said the people among whom I am number six hundred thousand men on foot, and yet You say I will give them meat and they will eat for a month! If flocks and herds be slain for them, will they be sufficient? Or if all the fish in the seas be gathered together for them, will they be enough?" (Num. 11:21, 22). The reply: "And the Eternal said, is the Eternal's hand short?" (Num. 11:23).

These persons had intimate and frequent interactions with the biblical Divinity. Yet the perception was then, as it is today, that there is nature and there is God, a dichotomy. It is difficult, perhaps impossible, to internalize intellectually the biblical concept that nature is just one manifest aspect of a Unity transcending all existence and therefore subservient to it.

God satisfied the Israelite yearning for meat by having a wind blow quails, exhausted from their flight over the sea, to the campsite. There they fell in great numbers (Num. 11:31). The *New York Times* reported on a similar event under the title "When Quails Come Back to Alexandria."[1] In this more modern version, restaurateurs gathered the birds as they fell on the shore, exhausted from their trans-Mediterranean flight. Nature does have the habit of repeating itself!

We have the burden of choosing between nature and divinity as the force behind the wind that three thousand years ago split the sea at the Exodus and later brought the quails.

The second key message that the Bible relates with regard to a divinely organized world is not to expect life to be a bed of roses without thorns. Thorns there are aplenty. Bad things are going to happen to good and to bad people. Cain murdered Abel. It was Abel whose sacrifice was accepted by God. If Abel was the good guy, why didn't God protect him? Two aspects of biblical religion arise from this event: bad things happen, and God lets them happen.

Job was a God-fearing good person. Why have him suffer such outrageous catastrophes? Abraham was told to leave his home and go to Canaan, the Promised Land. A mere four verses after Abraham arrived in Canaan, a famine struck (Gen. 12:10) that was so severe he had to leave Canaan and sojourn in Egypt. Famines also plagued his child, Isaac, and grandchild, Jacob. Later I search for a "why" to these sufferings. Here suffice it to say that when grief is dealt to those who seem least to deserve it, whether this grief arises from natural causes (famines, for example) or from human malice (Cain), this is no contradiction to the biblical concept of an omnipotent God. It may contradict our preconceived notions of what a God should allow or disallow, but confirmation of our notions is not what we are seeking.

Some gains, the Bible tells us, may be inherently associated with pain. Abraham is offered a deal by God: his progeny will be afflicted strangers in a land not theirs for four hundred years; in exchange, at the end of that period they will become a nation in their own land (Gen. 15: 13–14). Abraham could have replied as one might to a bee: I will forego your honey and your sting. But Abraham realized that forging a nation might require the pain of exile. He accepted the bargain. If we mistake pleasure to be the avoidance of pain, we may miss some of the greatest pleasures in life, such as reaching the peak of a mountain or rearing children. Ask parents the source of their greatest pleasure, and then ask them the source of their greatest pain. It's their kids every time.

In his closing address, Moses adjures the people to "Remember the days of old, consider the years of each generation" (Deut. 32:7). Kabalah tells us these "days of old" are the six days of Genesis, and "the years of each generation" are the historical records of civilization.[2] Understanding the events of our cosmic and social past is a key to discovering the immanence of God. The Bible insists the evidence is there for us to discover God in this world: "*You shall know that I am the Eternal*" (Ex. 6:7; Ex. 29:46; Deut. 4:39).

If studying history is indeed a path for all humanity to discover

God, then the reason for one of the more contentious and misunderstood issues of the Bible becomes clear. Having a people chosen to be "holy" becomes a necessity, both biblically and scientifically. The Hebrew word for holy is *kodesh,* which means separate, set apart.[3] As Balaam, the gentile prophet, stated, "It is a people that shall dwell apart" (Num. 23:9). The modern connotation of holy misses this, relating the word to being worthy of adoration. It is no wonder the Talmud likens the translation of the Hebrew Bible into other languages to a lion being locked in a cage, comparing it to the destruction of Solomon's Temple in Jerusalem.

In the language of experimental science, this "holy" people is an identifiable control group set apart against which the flow of history can be compared. As Nobel laureate and physicist Leon Lederman stated so pithily, "There is something spooky about the people of Israel returning to the land of Israel after 2,000 years." Had the Jews not been told that as a group they would return to the land after a horrific dispersion (Deut. 30:1–5), Lederman would have no basis for judging the reestablishment of the State of Israel as spooky. For spooky we might better substitute "teleological."

The best control is one that is present in the actual environment. The problem becomes how to maintain the separate identity of that people even while they are part of society in general. The Torah accomplishes this by presenting them with a list of constraints (foods, clothing, holidays). For three thousand years it has succeeded. To compensate for the burden of being set aside, those chosen to be separate needed a reward to offset the difficulty of the task. According to the Bible, that reward included a method, not necessarily unique or exclusive, to help in discovering and understanding the transcendental unity that forms the base of our universe.

Being "holy" does not mean being intrinsically better. It means being made visible as a symbol. A basic tenet of the Hebrew Bible is that all people have access to a higher reality. Isaiah stated this so clearly 2,700 years ago: "My house shall be called a house of prayer for all people" (Isa. 56:7). The God of the Bible is a God for all nations. Balak, king of Moav and an adversary of Israel, sent to Mesopotamia for a gentile prophet who came and prophesied (Num. 22–24). Jonah carried God's message to the foreign city of

Nineveh. Amos (Amos 9:7) reports that God took "the Philistines from Caftor and Aram from Kir."

Exodus, we learn from Amos, is an experience not confined to Israel alone. At least two other nations, Philistine and Aram, also experienced one. Perhaps they also were to be holy, but because they opted to neglect the message of their exodus those ancient nations with which God once spoke have disappeared. The choice was theirs. How an individual acts determines the clarity by which that individual realizes the unifying base of existence. As if to emphasize this, God tells the Israelites explicitly that their being chosen is not because they have inherently superior virtues as a people (Deut. 9:4–6).

The holiness of Israel is for the aggrandizement of God, not for the aggrandizement of Israel. This is seen in the texts already cited. At the golden calf (Ex. 32:14) and at the spies' rebellion (Num. 14:20), Moses argued with God that destroying Israel would make the nations of the world reject belief in God's omnipotence. If Israel's redemption were the sole goal of the Exodus, then the opinion of the other nations would be irrelevant. However, as the narrative of these passages makes blatantly clear, the opinion of the other nations is highly relevant. Biblically, the role of Israel is to be witness for a transcendent yet immanent Creator, active not only at the creation of the universe, but also in the universe throughout time.

In the unfolding of history, God's presence was perceived by many peoples in different ways. Based on the wonders of the Exodus from Egypt, Jethro, Moses' gentile father-in-law, realized that the "Eternal is greater than all other gods" (Ex. 18:11). Naaman, captain of a pagan nation's army, understood the biblical God even more clearly. Having been cured of leprosy through the instructions of Elisha the prophet, Naaman acknowledged that other than the Eternal "there is no god in all the Earth" (2 Kings 5:15). Rahab, a resident of Canaanite Jericho, hid the two scouts Joshua sent to reconnoiter Jericho. She did so because she found the Eternal is God "in the heavens above and on Earth below" (Joshua 2:11).

Moses alone grasped the full significance of the Unity, declaring "Hear Israel the Eternal our God the Eternal is One." With no equivocation, he stated: "You shall know today and place it in your heart that the Eternal he is God in the heavens above and on the earth below, *there is nothing else*" (Deut. 4:39). Moses had discovered that all existence is a manifestation of the Creator.

So the biblical God is one who gives leave to chance, who insists He will be seen in nature, who allows free will and injustice. But is He in control? The greatest point of contention between science and religion rises when believers insist God directly controls nature, while scientists insist that nature can run "on its own." Which is right?

According to the Bible, that divine control may be enigmatic. Even in those episodes when an omnipotent and active God is clearly in control, progress may not be smooth and direct. The most dramatic example of that is the Exodus.

God led the Israelites through the wilderness by a pillar of cloud in the day and a pillar of fire in the night. Cloud and fire, around the clock, inform us that God was leading every step of the way. As they marched directly toward their goal of Mount Sinai, God told the Israelites to change course, to turn around and head back along part of the path they had just traversed (Ex. 14:2). This placed the Israelites in great danger. The Gulf of Suez now blocked their escape from the pursuing Egyptians.

The reversal was not a divine afterthought. It was part of a plan with a multitude of goals, one of which was a test of the people's trust in God even though events might seem not to be proceeding as scheduled. The generalized lesson of the scenario: even with the infinite Guide in charge, reversals are part of the biblical system, be they in society or in nature. If we could see the entire scheme, we might know their reason (Ex. 14:3).

The biblical descriptions of the Creator interacting with the creation in a nonobvious fashion match scientific descriptions of our world tuned perfectly for life by some fortunate quirks of nature. While this does not prove there actually is a God, it does lay to rest those hollow claims that if indeed the purportedly infinite God of the Bible does exist, such a God would never have allowed an eye designed with a blind spot in so important a creature as the human, or have the dinosaurs appear only to destroy them 160 million years later. From the Bible's worldview, there is no reason to expect that the path from the primitive amoral aquatic animals of day five, 530 million years ago, to sentient, moral humans near the close of day six, 6,000 years ago, would be any more direct than was the trek of the Israelites from Egypt to Sinai.

In juxtaposition to skeptics' demands that an infinite biblical God would exhibit infinite and perfect control are the demands of the believer insisting that each species, from microscopic copepods to mammoth elephants, be a separate creation. An infinite God obviously could produce this variety of creatures, but this would not be the God described in the Bible. Throughout the text, we read descriptions of nature functioning naturally. Why the creationist paradigm insists on selling short the phenomenal laws of nature is beyond me. Atheists look in awe and wonder at the incredible precision by which the universe functioned to produce an Earth brimming with life. The fossil record, exhibiting the immediate appearance of life on the just-cooled Earth 3.8 billion years ago, the amazing explosion of life three billion years later, the several massive extinctions that redirected life's flow—most significantly 250 million and 65 million years ago—might superficially be seen as random events. Scrutiny of the complex and interrelated processes involved in those "days of old" makes this position barely tenable. The fossil and cosmological records may not be compatible with the view of God in Newton's day, but they are eminently compatible with the God of the Bible—a God clearly described as having created a universe and by that creation having instilled a leeway therein for chance in nature and choice by man.[4]

Of all the ancient accounts of creation, only that of Genesis has warranted a second reading by the scientific community. It alone records a sequence of events that approaches the scientific account of our cosmic origins.

The cosmogonies of the Greeks and Romans are filled with grotesque, violent, even obscene battles among warring gods, each acting as if he or she were the personification of a force in nature. Peals of thunder gave the Norwegians their Thor. The planet Jupiter's motion across the night sky was enough to convince the Romans that it must be the supreme deity. (They had hundreds of others as well.) Closer to home, biblically speaking, the Babylonian/Mesopotamian account of human ancestry puts into historical perspective the dramatic change wrought by the Bible.

According to the Babylonians, there was no material creation from nothing as envisioned by Genesis. Instead, from an eternal divine swamp, gods emerged, often in pairs. (Swamps are common along the banks of the Tigris and Euphrates rivers.) From the swamp gods emerged other gods for land, for sky, for the earth. The new young gods wanted their share of the glory and so started killing the older gods. Marduk, son of Ea, the earth god, appeared and did battle with Tiamat. Tiamat was one of the first of the gods to have gotten out of the swamp and so she had a large helping of primordial power. Being unable to kill her directly, Marduk caught her in a huge net and with a massive club, crushed her skull. He then cut her carcass in two and suspended one part to form the heavens. Not to be wasteful, Marduk mixed spilled divine blood with earth and formed humans from this godly mud. In the literal sense, man was a blend of heaven and earth.

Abraham, whom the Bible claims realized the concept of one ethereal, ethical God, creator of heaven and earth, grew up in Ur, a city in the heart of Mesopotamia. Considering the barbarity of the Tiamat/Marduk account, one can imagine the resistance Abraham's conceptual revolution must have met.

Upon this Babylonian tapestry, biblical religion wove a new concept. It was to transform humankind's understanding of self. While this does not prove its divinity, it does demonstrate its effectiveness.

To a world steeped in the barbarity of child sacrifice and multiple gods, each bribable and incestuous, the book of Genesis revealed a world never before envisioned. There was order, and there was the possibility of progress. The genesis described in the first chapter of the Bible broke the pessimism of the pagan world. The belief in unending cycles with no opportunity to rise was gone. Here was a conceptual revolution, the gradual transformation from a world "unformed and void" (Gen. 1:2) to one filled with a celebration of life. Egypt's Book of Death had been replaced by a Tree of Life (Prov. 3:18). No longer did one part of nature have a power that could coerce another. Prayer and sacrifices were to induce a change in the offerer, not in the deity. Truly, evening had passed and morning had come.

The barbarity of Rome persisted for a while. As did the perverted democracy of Greece which granted civil rights to the tiny landed fraction of the population while holding all others in perpetual slavery. Parts of their cultures were incorporated into other religions. But the dawn had irrevocably arrived. A light to the nations was shining (Is. 49:6) and its source was recorded in a way that has excited children and intrigued adults for the three thousand years of its history.

The demand to love our neighbors (Lev. 19:18) and the formula to achieve that love, abandon grudges, abandon vengeance (ibid.), the extraordinary concept of one law for both stranger and home-born (the need to respect a stranger is repeated thirty-six times in the Five Books of Moses, more often than admonitions concerning the sabbath), these and a host of other civil guides found in the Five Books of Moses shaped Western society. And all rest on the foundation defined in the one sentence considered by Jew and Christian (Matthew 22:37; Mark 12:29) to be the most important statement ever made: "Hear Israel, the Eternal our God, the Eternal is One" (Deut. 6:4). Oneness is the name, the essence, and the mark of the biblical God.

A landmark had been set for all time: the revelation that beneath the multifaceted complexity of our lives lies a unity binding not only all humanity, but all existence. That we are made of stardust is a truth surpassing poetic license. We are literally children of the universe.

We have searched out the Bible's description of God's imprint in our universe and found an interaction similar to a parent allowing a child the freedom to learn independently with occasional guidance when actions get too far off course. The creation for the most part will look as if an unguided nature is in control. To discover the Divine we must take more than a superficial glance at the world around us.

"And those who trust in You know Your name" (Ps. 9:11).

CHAPTER 6

Life:

Its Origins

and

Its Evolution

However improbable we regard this event [the start of all life], or any of the steps which it involves, given enough time it will almost certainly happen at least once. And for life as we know it . . . once may be enough.

Time is in fact the hero of the plot. The time with which we have to deal is of the order of two billion years. What we regard as impossible on the basis of human experience is meaningless here. Given so much time the "impossible" becomes the possible, the possible probable, and the probable virtually certain. One has only to wait: time itself performs the miracles.[1]

These words were written by Nobel laureate and Harvard University biology professor George Wald and published in the widely read journal *Scientific American.*

For decades leading biologists had promulgated the position, stated so well by Wald, that time and chance were the forces behind the miracle of life. It was logically correct. After all, what else could be operating?

Wald's definitive statement, made on behalf of the scientific community, rested firmly on research completed the previous year. In 1953, Stanley Miller, then a graduate student at the University of Chicago, had produced amino acids by a series of totally random reactions. His experiment was simple but brilliant.

Miller evacuated a glass flask and then filled it with the gases thought to have been present in Earth's atmosphere 3.8 billion years ago: ammonia, methane, hydrogen, and water vapor. Free oxygen was not present. It appeared only billions of years later, the product of life itself: photosynthesis. Using electrodes placed through the walls of the flask, Miller discharged electric sparks, simulating lightning, into the gases. Their energy induced random chemical reactions among the gases. After a few days, a reddish slime appeared on the inner walls of the apparatus. Upon analysis, the slime was found to contain amino acids.

The importance of Miller's experiment was at once apparent. Amino acids are the building blocks of proteins and proteins are the building blocks of life. As Wald pointed out, two billion years had passed between the appearance of water on Earth and the appearance of life. If random reactions in a small flask can produce amino acids in just two days, given two billion years of reactions throughout the Earth's vast atmosphere and oceans, the first forms of life, bacteria and algae, must have been the product of similar random reactions during those eons. The impossible had become the probable and the probable certain. We and all other members of the biosphere are living proof of the theory's accuracy.

The news media worldwide reported the significance of Miller's seminal experiment. The public had been told the truth: life had started by chance.

Or had it?

Wald's article was such an important statement that twenty-five years later, in 1979, *Scientific American* reprinted it in a special publication titled *Life: Origin and Evolution*. The only difference was that this time it appeared with a *retraction*. I have seen no other retraction by a journal of a Nobel laureate's writings.

The retraction was unequivocal:

> *Although stimulating, this article probably represents one of the very few times in his professional life when Wald has been wrong. Examine*

his main thesis and see. Can we really form a biological cell by waiting for chance combinations of organic compounds? Harold Morowitz, in his book "Energy Flow and Biology," computed that merely to create a bacterium would require more time than the Universe might ever see if chance combinations of its molecules were the only driving force.[2]

In short, life could not have started by chance.

Lest you think that the scientific community has changed its opinion since 1979, the following appeared in the same journal in February 1991, in a review article by John Horgan on the origins of life: "Some scientists have argued that, given enough time, even apparently miraculous events become possible—such as the spontaneous emergence of a single-cell organism from random couplings of chemicals. Sir Fred Hoyle, the British astronomer, has said such an occurrence is about as likely as the assemblage of a 747 by a tornado whirling through a junkyard. Most researchers agree with Hoyle on this point."

Since 1979, articles based on the premise that life arose through chance random reactions over billions of years are not accepted in reputable journals.

This is, of course, not an affirmation of the existence of a Creator. However, it does assert that the simplest forms of life, single-celled bacteria and algae—and even viruses, which are one step below life—are far too complex to have originated without there being an inherent chemical property of molecular self-organization and/or reaction-enhancing catalysts at every step of their development.[3-5] These conditions must have been present to jump-start life on Earth.

The Bible has no problem with this concept of life's beginnings: "And the Earth brought forth" life (Genesis 1:12). In biblical language we are being told that the Earth itself had within it the properties to encourage the emergence of life. There is no biblical mention of a special creation for the origin of life. The laws of nature, created along with the creation of the universe, and the very special conditions of the Earth were quite adequate to orchestrate the flow of the universe toward life.

Articles authored by Nobel laureates are not lightly retracted. The statistical computations by Morowitz may have cast a shadow of doubt over Wald's claims for the power of chance, but I question whether *Scientific American* would have actually retracted the article based on statistical calculations alone. The editors were not flooded with letters disagreeing with Miller's and Wald's thesis on the random origin of life. Scientific opinion of the day was that life had started via a series of chance random reactions.

The article was withdrawn because research performed by another Harvard professor proved Wald wrong. In the 1970s, Elso Barghoorn, a paleontologist, discovered micro-fossils of bacteria and algae in rocks close to 3.5 billion years old. Deposits representative of organic carbon appear in formations 3.8 billion years old. That is also when the first liquid water appeared on Earth, and hence the first time that life could survive. All life on Earth is water based. No water, no life, but with water, life was possible. It had only to develop, and develop it did, *immediately* in the presence of water. There were no "billions of years" for the amino acids to combine randomly into life.

So suddenly did life arise on Earth that, the theoretical biologist Francis Crick wrote, "Given the weaknesses of all theories of terrestrial genesis [the origin of life on Earth], directed panspermia [the deliberate planting of life on Earth] should still be considered a serious possibility."[6] Crick certainly understands the complexity of life. With Watson and Wilkins, he shared the Nobel prize for discovering the shape and functioning of DNA, the genetic code of all life.

Perhaps it is merely coincidence that Crick and his associates published their research on the complexity of life in 1953, the same year that Stanley Miller had "proven" the simplicity of life's origin.

Crick suggests the concept of the "directed" planting of life on Earth. He is a strong enough mathematician to realize that a random sprinkling of the universe with the seeds of life, by a sort of cosmic Johnny Appleseed, gives slim chance to those seeds ever arriving at a hospitable planet. The demands of life are stringent and the universe is for the most part quite hostile. A cosmic seeder of life would have to aim at a particularly proper planet.

Do we hear in Crick's words the hint of a godly Guide in the development of life? Not on your life! He maintains he is an agnostic with a prejudice toward atheism.[6]

As of this writing, science has no agreed-upon explanation for the cause. It is an important (and currently undersupported) area of research. But whatever theories are put forth, the truly extraordinary fact remains: as soon as the conditions on Earth arose for life to exist, life appeared.

What about the flow from those first simple one-celled forms of life to the complexity of nature that we see today?

The difference between secular evolutionists and theologians is not in the details of the events. The difference is that the former claim the development is all by random mutations while the latter see a channeling in the flow of life that implies a teleology.

To distinguish between direction and randomness, we must study the flow of life in detail.

There is a popular impression that fossils have proven the validity of classical evolution. Yet most paleontologists admit this is not the case. According to the picture presented by the fossil record, bursts of morphological change occurred within startlingly brief periods of time. This staccato aspect of the fossil record had not been predicted. Its discovery has called for basic rethinking concerning the mechanisms that drive evolution to ever greater complexity. These rapid changes cannot be explained by purely random mutations at the molecular genetic level. In light of the mounting evidence that the classical concept of evolution is flawed, the journal *Science* featured a peer-reviewed report titled "Did Darwin Get It All Right?" In that article we learn that "the most thorough study of species formation in the fossil record confirms that new species appear with a most un-Darwinian abruptness."[7]

The burst of multicellular life at the start of the Cambrian, 530 million years ago, was so dramatic that the *New York Times* reported on it under the page-wide, 2-cm-high headline, "Spectacular Fossils Record Early Riot of Creation."[8]

Dr. Jan Bergstrom, the paleontologist cited by the *Times*, suggested that for new morphologies to have developed this rapidly, "you could have the formation of an entirely new type of animal within thousands of years." Considering the complexity of the DNA genetic codes which shape all life, the chance formation of a "new type of animal within thousands of years" requires an active imagination. It strains the credibility of random molecular evolution. Perhaps that explains the choice of the word "Creation" in the headline.

The scenario I learned in school was quite different: One hundred to two hundred million years were said to have passed as invertebrates gradually evolved into vertebrates. First came amorphous creatures such as sponges, then as more complexities were added worm-like annelids evolved, then molluscs, and finally, after a host of other evolutions, on to vertebrates.[9] (See Figure 4.)

The fossil evidence that challenges this classic concept of evolution has been found worldwide: in western Canada, near Chengjiang in southern China, in Africa, Greenland, and Sweden. The Cambrian explosion of life encompassed the globe. Jointed legs, food-gathering

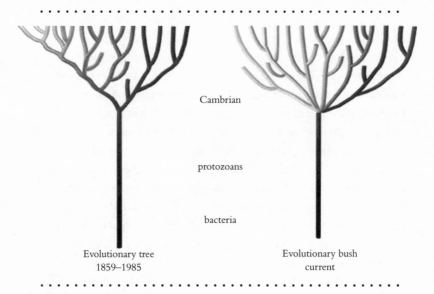

Cambrian

protozoans

bacteria

Evolutionary tree   Evolutionary bush
1859–1985   current

FIGURE 4: Two concepts of evolution

Until the mid-1980s the understanding of the development of animal life was that it had followed the logical path of a gradual evolution with more simple phyla over eons leading into more complex phyla. With the rediscovery of fossils held quietly in the drawers of the Smithsonian Institution since 1909, this concept underwent a drastic revision. These fossils in conjunction with other discoveries indicate that all animal phyla appeared almost simultaneously 530 million years ago in the Cambrian period. All further development was confined to variations within each phylum. One of the great mysteries of animal evolution is why no new phyla have appeared since that Cambrian explosion of life.

appendages, intestinal structures, notochords, gills, eyes with optically perfect lenses—all these "evolved" simultaneously. Sponges, rotifers, annelids, arthropods, primitive fish, and all the other body plans represented in the thirty-four animal phyla extant today appear as a single burst in the fossil record. And it happened 530 million years ago. Those are the data. No one disputes them.

Based on radioactive dating of rocks that bracket the Cambrian explosion, the development occurred within a period of five million years.[10] The sediment deposits of this five-million-year span are at places 300 meters thick. Throughout the depth of this sediment, and therefore over the entire five million years, there is little or no change in the morphologies of these animals.[11] In a leap, life moved from single-celled protozoa and the amorphous Ediacaran clumps to multicellular complexity. According to the fossil record as we currently know it, the simultaneity was literally true.

This burst of life had been explained away as an artifact of the process of fossilization. If only hard tissues form fossils, then evolution of soft-bodied animals will go unrecorded. Popular wisdom had it that prior to the Cambrian, animal life must have evolved as soft-bodied. With the Cambrian came life's discovery of bone and shell. Hence, these would logically be the first forms of multicellular life that could form fossils. Of course this brushed aside the knowledge that many Cambrian fossils were of soft-bodied creatures and that fossils of "soft" bacteria and protozoans predate the Cambrian by three billion years.

The soft-body theory was destroyed by a nasty fact (a phrase ironically attributed to T. H. Huxley). Until the early 1990s, the fossil record for the one hundred million years prior to the Cambrian explosion was mostly a blank with regard to multicellular animals. The assumption was that evolution had occurred gradually during this period, but that fossils had not been formed or at least not yet found. With the discoveries of Ediacaran fossils from this period in Namibia, the British Isles, and elsewhere, this hope faded.[12,13] The fossils of this formerly blank period show no progress beyond globular pancake-like shapes lacking appendages and mouths, some having small portions of shells, but most appearing to have been completely soft-bodied. The Ediacaran fossils found in pre-Cambrian strata worldwide are simple and soft-bodied.

So how can the development of life be explained? Over three billion years spanned the gap between the immediate appearance of one-celled life on the just-cooled Earth, 3.8 billion years ago, and the explosion of multicellular life 530 million years ago. Perhaps during those eons, random mutations in the one-celled antecedents of all modern life stored genetic material containing latent (neutral) information potentially useful for the impending explosion of animal complexity. This would then be available for expression suddenly and simultaneously at the inception of multicellular life. Species of modern algae and protozoans have the space in their DNA for this neutral information. Each of their cells contains as much as one hundred times more DNA than a cell of any mammal, including humans.[14] Since micro-fossils of primordial algae and protozoans have shape and size similar to modern specimens, it may be inferred that their genetic library was equally large.

We know that the genetic material of many plants and animals contains blocks of latent information able to be immediately expressed by changes in the environment or by single mutations. Chickens are known to grow hair, not feathers. Horses are born with multiple digits.[15] Human babies unfortunately at times emerge from the womb with their "gill slits" still open. Both mammals and birds produce gill arches during embryo development, but then both mammals and birds, according to the fossil record, share a common origin as primitive fish. Marsh plants when submerged in water develop a leaf structure different from that of the same plant grown on dry land, even when cloned and therefore genetically identical.

According to the "latent library" theory, all this information is quietly present in the genome, waiting for the cue to be expressed. When the marine lizard ichthyosaurus appears suddenly in the fossil record with an outer shape essentially that of a fish, or a land mammal becomes fish-like in shape as theorized for the whale, we may be witnessing the historic shift of preexisting genes from latent to active states. Fossils of what appear to be ancient primitive whales having small vestigial hind legs have been found in India and Pakistan.[16,17] If the fossil record is correct, the phylum Chordata, of which both lizard and mammal are members, has at its base primitive fish. Some fish-like genes are certainly held in the genomes of modern land-based chordates, hence the occasional gill slits of human babies and the gill arches and yolk sacs of mammal embryos. Reorganization and expression of stored genes could account for a rapid "evolution" of the fish-like characteristics of these species.

The tunicate, or sea squirt, has a larval stage that evolves toward a form similar to primitive fish of the Cambrian period. Its larval eye has lens, retina, pigment, and cornea. All of these are gone in the adult tunicate, as is the fish-like shape. But the genes remain. Loss of the structure need not mean loss of the ability to form the structure at some other time, in some other environment.

The obvious questions with regard to algal and protozoan genome size are: why does an algal cell or an amoeba retain so much genetic capacity? And why, within this huge genetic library space, would a primordial protozoan have stored information related, for example, to jointed limbs or vertebrae? There would be no immediate benefit to the amoeba and so no genetic reason for them to maintain this neutral information in their DNA. If it were not neutral, then its expression was clearly not in the forms as expressed in the Cambrian animals.

The concept of a latent library posits a mechanism very different from the classical theory of evolution wherein random mutations provide the changes in morphology. Nonetheless, primordial preprogramming of life's developments is exemplified in the morphogenesis of the eye. A gene group, Pax-6, is a key regulator in the development of eyes in all vertebrates. Its analog (a very similar gene) has been found to control development of the visual systems of molluscs, insects, flatworms, and nemerteans (ribbon worms). These represent five of the six phyla that have visual systems. (The sixth phyla with vision has not yet been studied.) The molecular similarity among these analogs is nothing less than astounding.[18,19] The paired domain of the gene contains 130 amino acids. The match of these amino acids between insects and humans is 94 percent! Between zebra fish and human the match is 97 percent.

Could five genetically separate phyla have evolved these similar genes individually by chance? There are twenty different amino acids available to fill each of the 130 spaces on the gene. This means there are 20^{130} or 10^{170} possible combinations. There are one hundred million billion billion billion billion billion billion billion billion billion billion billion billion billion billion billion billion billion billion ways the amino acids can arrange themselves in those 130 slots on the gene. The number far exceeds the number of particles in the entire universe.

Any combination is possible one time. Getting the same or even similar combinations a second time is the statistical problem. The likelihood that random mutations would produce the same combination five times is 10^{170} raised to the fifth power.

There is no way this same gene could have evolved independently in each of the five phyla—it must have been present in a common ancestor. The gene that controls the development of eyes was programmed into life at the level below the Cambrian. That level is either the amorphous sponge-like Ediacarans or one-celled protozoa. But *neither has eyes.*

Evolution is not a free agent.[20] The laws of biology, chemistry, and physics, the laws of nature, determine which structures can evolve. The thirty-four basic body plans of all animals extant since the inception of animal life are the only body plans that fit within those constraints of nature. We can only conclude that evolution is channeled along these paths, working new variations into old original themes.

Morphological constraint is evident not only at the level of individual genes. It is also seen in complex body organs when *very different* phyla of animals develop very *similar* organs to satisfy similar needs. Being of different phyla, they have been genetically separated since the inception of multicellular life. Hence, they developed these similar solutions *independently.* Is it simply logical that similar biological needs will be solved by similar solutions? More likely, these separated-yet-similar convergent organs again demonstrate that there are not a myriad of choices in the grab-bag of evolution. Inherent in nature are channels providing direction to the development of life.

Sight is a wonderful sense, one of the great joys in life. The design of the human eye is so successful for granting sight that all vertebrates have the same basic organ that allows us humans to marvel at the beauty of a sunrise. But then this ocular similarity among vertebrates is not surprising. All vertebrates are similar in much of their bodily structure. Vertebrates have been genetically connected for 530 million years, since the first appearance of multicellular life. What is surprising is the similarity of the human eye to the eyes of several types of animals that fall within the phylum of Mollusca. The best known example is the eye of the octopus, a member of that phylum.

Mollusca and vertebrates, as with all the thirty-four animal phyla, separated 530 million years ago, at the developmental level just above protozoa. Yet their solutions to the need for sight are almost identical. Although the choice of proteins is different, both have eyes with cornea, functioning iris, lens, vitreous humor, three-layered retina with rods, pigments for moderating light intensity to the retina, a ganglion of nerves connecting the retinal photoreceptors to the brain, and most subtly of all, lateral inhibition, the mechanism by which adjacent optic neurons interact in order to enhance the perception of boundaries between two similar colors.

Actually, the octopus eye has an advantage. Its retinal nerves leave directly from the back of the retina. The mammalian eye has the nerves in front of the retina; they must then pass through the retina to reach the brain. This passage produces a millimeter diameter puncture in the retina resulting in the well-known blind spot in our vision. The difference in the "wiring" of the eye-brain connection arises from the difference in the development pattern of these organs. The octopus retina forms from optically sensitive skin cells. The vertebrate retina forms from optically sensitive brain cells.

We refer to this remarkable similarity of organs in very different animals as convergent evolution. But can we explain it? The lottery of individual random mutations at the molecular genetic level cannot be its only driving force. A simple calculation can show that the likelihood of producing any particular sonnet of Shakespeare by random typing is about one chance in 26^{490} or one in 10^{690}. Here we are asking for chance to produce the same "sonnet" twice. The statistical improbability of pure chance yielding even the simplest forms of life has made a mockery of the theory that random choice alone gave us the biosphere we see.

"The hypothesis that the eye of the cephalopods [i.e., the octopus and squid] has evolved by convergence with the vertebrate eye is challenged by our recent findings of the [human] Pax-6 related gene sequences in octopus and squid."[21] That is a quote from *Science,* a leading journal. In technical language it says that the astounding similarity between octopus, squid, and human eyes makes it statistically so improbable as to be functionally impossible that they evolved from different origins and "converged." Convergent evolution does not occur by chance. It cannot have occurred by chance.

Evidence from anatomy, from molecular biology, and from the fossil record is that evolution is channeled in particular directions. In that sense, we are written into the scheme.

EVOLUTION AND THE FOSSIL RECORD

Animals first appear on day five of Genesis when we are told the waters swarmed with animal life (Gen. 1:20). Genesis also records the subsequent appearance on day five of a group of animals referred to as *oaf*. This Hebrew word, often mistranslated as "bird," means a winged animal. The Hebrew word for bird is *tsepoor* (Gen. 4:7). Current theory relates the first appearance of wings to water insects, some 330 million years ago.[22,23]

There is no fossil evidence of primitive wings prior to the appearance of fully developed winged insects 330 million years ago. It is hard to explain the sudden leap from winglessness to wings with a 30-cm span. That is quite a mosquito. But those are the data as revealed by the fossils. A butterfly emerges from a caterpillar because the DNA of the caterpillar contains the information of the butterfly's wings. Were wings preprogrammed in the earlier insects? That would account for their sudden appearance.

The staccato nature of the fossil record has bothered paleontologists for decades. This problem is magnified when we recall that it is not a mutation in any random cell that is required to produce an evolutionary change. The mutated DNA must occur in the cells responsible for the production of progeny, the eggs of the female and the sperm-producing germ cell of the male. And then, if the mutation is in the DNA of an egg, that particular egg must be fertilized and survive to become a progeny-producing adult.

How have evolutionists explained the fossil record? Gradualism fails—so rapidity has become an important part of evolutionary theory. A rapid change will not likely leave many or possibly any fossils to record the morphological flow from the old to the new type of animal. The new, more fit species will compete for natural resources more effectively and in "short" time will destroy the old species. Millennia later, so the theory goes, paleontologists will find a fossil record of the new morph directly above the old morph with no transitional link which might embody an animal that was part the older and part the newer species.

In the entire fossil record, with its millions of specimens, no midway transitional fossil has been found at the basic levels of phylum or class, no trace of an animal that was half the predecessor and half the successor.

Well, that's not entirely true. One such fossil has been found. To some theorists, it's the centerpiece in the proof of evolution, described as the Rosetta stone among fossils.[24] That fossil, the archaeopteryx, demonstrates to evolutionists that a line of reptiles developed into birds. The specimen at the British Natural History Museum is the only fossil of the museum's myriads of fossils that is locked separately.[25] Such is its importance. Six specimens have been discovered between 1861 and 1987, all in one region of western Germany.

Some persons claim the archaeopteryx is a forgery,[26] a flying version of the Piltdown man. They suggest that a bird and a lizard have been pressed together into the stone as an attempt to prove evolution correct and the Bible wrong. As a student of the Hebrew Bible, I'm all for the validity of the archaeopteryx.

THE LINK THAT NEVER WAS REALLY MISSING

The archaeopteryx lived 150 million years ago, in the late Jurassic period, the era when reptiles dominated the main ecological niches. At the time, central Europe, including what is now Germany, had a tropical climate. The specimens of the archaeopteryx appear to have fallen into a saltwater lagoon and been covered with silt. Their preservation by nature was so gentle that the structure of their feathers has been retained in the fossil.

Feathers on wings speak of a bird. So does the archaeopteryx's furculum or wishbone. The wishbone arises from the fusion of the two collarbones and provides the necessary solid base for the wing muscles. The feathered tail is further evidence that the archaeopteryx was a bird. But this animal also had jaws with teeth, not a beak, a long bony tail, and claws on its wing-like arms and its feet. It begins to sound like a reptile and not a bird.

Or was it a mix of the two? It is called "the perfect example of a transitional form in the evolution of modern birds from reptiles."[26] It combines two distinct classes (reptile and bird) in a single animal. So impressive is this link in the line of evolution that "a visiting professor actually fell upon his knees in awe" when he first saw the specimen housed in Britain's Natural History Museum.[27]

In the third book of the Bible, Leviticus, there is a list of ritually pure and impure animals. The list divides animals into categories: the insects in one place, fish in another, and so forth. In Leviticus 11:18 birds are listed. Among them we find the *tinshemet*. Twelve verses later (Lev. 11:30), the reptiles are listed. And behold, the *tinshemet* appears again. The same name, spelled identically (*tuf nun shin mem tuf* in the Hebrew) is given for a bird and for a reptile because at one level of biblical meaning the animal fell into both categories.

In the entire Bible, there is the one reference to an animal that falls into two categories, the *tinshemet*. In the entire fossil record there is one fossil that falls exactly midway between two classes of animals, the archaeopteryx. And both the archaeopteryx and the *tinshemet* are part reptile, part bird. It is the "link" that never was missing.

The Greek and hieroglyphic inscriptions held so long on the black basalt of the Rosetta stone allowed us finally to decipher Egyptian hieroglyphics. Perhaps the archaeopteryx/*tinshemet* will provide an equivalent insight into our understanding of the biblical and paleontological descriptions of life's development.

The flow from the big bang to the abundance and diversity of life we find today in every nook and cranny of Earth has been a passage moving upstream against the relentless tendency of nature to proceed from order to disorder. The second law of thermodynamics tells us that all nonmanaged, or random, systems always pass to a state of greater disorder. That is why a cup of tea gets cooler when put aside during a twenty-minute phone conversation. The orderly concentration of heat within the cup becomes randomly dispersed among the molecules of air in the room. And when the cup falls and breaks, its pieces never reassemble into the cup it once was. Disorder is the statistical trend of nature simply because for any given collection of atoms, whether from a cup or a carp, the number of disorderly combinations is vastly greater than the number of orderly combinations. Unless order is imposed on a system, random choice among all possibilities always falls in that immense expanse we refer to as disorder.

The condition of the universe at its creation was one of chaotic, exquisitely hot energy. In the terms of physics, it was a plasma. Since the natural direction of nondirected, random systems is to degenerate into ever greater chaos, the universe might have expanded, producing an ever cooler random mix of energy and matter. Yet from the original chaos, there developed the symphony of life. The flow from the primordial state of universal chaos to the simplest forms of life and then on to humanity represents the imposition of order on that primordial chaos, an imposition somehow accomplished through the laws of nature created at the beginning.

This move toward order from chaos is not impossible *provided the system had direction*. There was and still is usable energy present to power the emergence of order. Order out of chaos is, however, such an unusual and improbable trend of events that the Torah mentions it six times.

The seemingly simple language of "And there was evening and there was morning," marks the closing of each of the six days of Genesis. That we must look for a deeper meaning in this sixfold repetition is obvious once we consider that the Sun appears on day four. How are we to understand the concept of evening and morning for those pre-Sun days? The kabalist Nahmanides explained it for us almost eight hundred years ago:[28]

The Hebrew word for evening is *erev*. The root of *erev* is disorder, mixture, chaos. The Hebrew word for morning is *boker,* its root being orderly, able to be discerned. In the subtle language of evening and morning, centuries before the Greek words of chaos and cosmos were ever written (Hebrew writing predates the Greek), the Bible described a step-by-step flow from disorder *(erev)* to order *(boker);* from the plasma of the big bang to the harmony of life.

Is the Divine active in this flow, but hidden and so explainable as just another act of "mother nature?" Is God immanent? I would not attempt to *prove* a case for divine direction. But there is a certainty:

twentieth century science has opened the door, and opened it wide, for that interpretation!

Sixty-five million years ago, the reign of dinosaurs over the animal kingdom was about to come to an abrupt end. It had lasted for 150 million years. The time was the close of the Cretaceous period.

The continents in their slow drift across the face of the globe had by then taken much of the shape they have today. Global temperatures were warmer. Less water was held as ice in the polar caps and so sea level was higher. Low-lying areas which in our time are the fertile midlands of the United States were then the beds of shallow seas. Thirty-foot-long parasaurolophus browsed beside the smaller sauropelta, feeding on rich vegetation. Herds of the three-horned triceratops moved along a river bed, ever wary of the threat from the meat-eating tyrannosaurus.

But a killer more lethal than the largest of the tyrannosaurus was about to strike. Its target was the Caribbean. Its prey, half the species of all life on the globe. A small (in cosmic terms) piece of debris left over from the formation of the solar system five billion years ago, a chunk of a supernova 10 kilometers in diameter not caught up during the aggregation of the other chunks of star dust that led to the making of the Sun, the planets, and us was hurtling toward the Earth at 30 km per second. No creature could see it. Its illuminated side was toward the sun. The space-black side faced the Earth. In the interplanetary vacuum of the solar system there are only two or three atoms per cubic meter. This is not nearly enough to produce the frictional glow of a meteor.

To feel the scale of this asteroid, think of looking off toward the horizon and seeing a more or less spherical jagged rock, the bottom of which is just touching the Earth's surface while the top is higher than the highest airplane you have ever seen. Now start walking toward this monolith. Soon it occupies your entire field of vision and it is still miles away. Such a rock was about to decimate Earth's biosphere.

The kill was not gradual. In less than two seconds the asteroid had punched a hole through the atmosphere and exploded into Earth, producing a crater 150 km in diameter and a few kilometers deep. All the energy of its speed was converted into heat, a blast fifty thousand times more powerful than the world's entire nuclear arsenal. The seismic wave was felt worldwide. Massive volcanoes erupted. Halfway around the globe, lava jolted free from the depths of Earth, covering all of northern India with a kilometer-thick blanket of the hot stone. Dust and vapor enveloped the planet in a shroud of darkness. Six months later, the noontime Sun was still dimmer than a quarter moon. Temperatures plummeted. Photosynthesis stopped, and all animals larger than five kilograms died. Mammals the size of squirrels had coexisted with dinosaurs for 150 million years, but the dinosaurs had controlled the ecosystem. Those clever, warm-blooded mammals survived that spoiler from outer space. The dinosaurs did not. Thus began the mammalian era in which we live.

This may have been an utterly random event—a pointless but significant collision. Alternatively, we can see the event as a divine retuning of the world. A force from outer space rechanneled the flow of life. Perhaps the dinosaurs were getting bigger but not getting smarter. The asteroid gave life the chance to redirect toward the desired goal of a sentient, intelligent being able to absorb within it the amazing concept of ethical monotheism. Those are always our options of interpretation, just as they were at the Exodus 3,300 years ago.

Moses and the Israelites left Egypt following a pillar of cloud by day and a pillar of fire by night (Ex. 13:21). Their exodus led first toward the coastal plain. That would take them around the Gulf of Suez. Then they were told by God to double back, to avoid the warrior Philistines who controlled the Mediterranean coast. The Israelites now headed toward the Sea of Reeds (Ex. 13:17, 18).

This is one of the great biblical acts of faith. The divine pillars of cloud and fire were leading them directly toward the sea! And still they followed. On the face of it, they were walking into a trap. The Egyptians were behind them in hot pursuit and the forbidding sea lay in front.

What occurred is stated explicitly in the book of Exodus. When the Israelites cried to God at the shore of the sea, a strong east wind started and blew all night (Ex. 14:21). The wind dried the seabed.

Why a wind, and why a wind that took all night to dry the seabed? If the divine plan was to use a wind, then why not a spectacular wind, something miraculous that would complete its drying task in seconds, a tornado or cyclone? Then all the world would know it was a miracle with no question.

Of course, the natural appearance of the wind was exactly the intent. Choices had to be made. For the Israelites, to trust in the Divine or to surrender to the Egyptians? For the Egyptians, to follow the Israelites onto the seabed or to retreat? Had the wind been obviously supernatural, the decisions would have been predictable, and free will would have been compromised.

The wind had such a natural feel to it that the Egyptians followed the Israelites right onto the just dried seabed (Ex. 14:23).[29,30] It was in fact such a seemingly natural wind that we had to be told of its divine source: "*The Eternal caused* the sea to go back by a strong east wind all the night" (Ex.14.21). Only when the Egyptians were trapped in the seabed did they acknowledge the miracle: "The Eternal fights for them [the Israelites]" (Ex. 14:25). The delay in their recognition of the Divine in the events of the Exodus proved to be fatal: "And the sea returned to its strength . . . and the Eternal overthrew Egypt in the midst of the sea" (Ex. 14:27).

The biblical message: not every extraordinary event in nature is labeled "miracle, made in heaven." Sometimes we must read between the lines to apprehend its full significance.

CHAPTER 7

Evolution:
Statistics
Versus
Random
Mutations

Statistical analyses of evolution are fraught with assumptions.
Because the events being studied occurred in the distant past, most
of these assumptions are not open to verification. Rates at which the
mutations occurred, the order of the mutations, the original DNA
information upon which the evolutionary mutations were imposed,
the environmental challenges during the periods of the change are
all unknowns. All affect the model being analyzed.

In addition, a conceptual error is frequently encountered. Questions are often asked about the probability that a specific animal or
organ could have evolved by random processes. Could a flatform or
a mollusc or a monkey have evolved from a protozoan within the
years indicated by the fossil record? The error of these questions lies
in the assumption that evolution was seeking a specific goal. Basic to
the entire concept of evolution is that no goal exists. If flatworms
had not emerged from the evolutionary cauldron, then roundworms,
or squareworms, or perhaps no worms but an altogether different
animal would populate that ecological niche today.

The lack of goal limits an analysis of evolution to calculating the
likelihood that the number (not the type) of mutations required for

the changes observed in the fossil record might have occurred within the time period indicated by the fossil record.

Life can be seen as a combination of proteins that have a symbiotic coexistence. We and all other members of our biosphere fit this description. But not all combinations of proteins function in a mutually supportive way. That is why animal life is confined to some thirty-four basic body plans (phyla). The Cambrian explosion produced other body plans too, but they failed to fulfill the requirements for survival. For a life form to persist in this environment we call our home, a very special combination of proteins is required.

Humans have approximately seventy thousand genes (some researchers say fifty thousand). All mammals have a similar number though not necessarily the same genes. Seventy thousand means approximately seventy thousand proteins. Proteins are strings and coils of between two hundred and a thousand amino acids. The seventy thousand genes would then be organizing some seventy million (70,000 x 1,000) amino acids into specific structures. These make up the thirty trillion cells of a healthy human. Can this arrangement be the result of random selection? Let's look at a few numbers.

First we notice that random generation of letters by a computer, or by the hypothetical monkeys typing away on typewriters, never produces meaningful sentences more than a few words in length. That's simply because the possible number of meaningless combinations vastly exceeds the number of possible meaningful sentences. With one hundred letters in a sentence, there are 26^{100} combinations of those letters. If one letter were written on each fundamental particle in the entire universe, it would take literally billions of billions of universes to complete the task of printing out the "text."

There are approximately 10^{18} bits of information in all the libraries of the world.[1] If each of these bits were contained in a one-hundred-letter sentence they would require 10^{20} letters. There is essentially zero chance (actually in the order of one chance out of 10^{120}) that any one of those sentences from all the libraries of the world would be generated by random typing. Randomness just doesn't cut it when it comes to generating meaningful order out of chaos. Direction is required. Always.

Within the thirty-four phyla that define the basic structure of all animal life, there are approximately thirty million species.[2] If they all had genomes and genetic information as extensive as humans (70,000 genes), and no proteins were common among species, then life would be constructed of two million million (2×10^{12}) different proteins. Take a model protein to be three hundred amino acids in length. There are twenty different amino acids used in life. The number of possible combinations of those three hundred amino acids is 20^{300} or in the ten-based system, 10^{390}. Of the possible 10^{390} protein combinations, life would only be using one in 10^{378}.

Why, when in excess of 90 percent of marine life died 250 million years ago, did no new phyla emerge? None emerged because none fit the requirements of life. Life has selected from the 10^{390} possible combinations the fewer than 2×10^{12} that work. The selection of that minuscule fraction (the one out of 10^{378}) of protein combinations that function for life from the vast number of possible combinations cannot have been by random point mutations on the DNA of the genome. It would be as if nature chose by random from a bag containing a billion billion billion . . . (repeated forty times) proteins the one that worked, and then repeated the trick a trillion times! If protein generation were a random process, then as with random word generation, the result would also be gibberish, but with life it would be fatal gibberish.

The only way random letter generation has a prayer of producing meaningful sentences is if the programmer instructs the computer how to recognize meaningfulness and how to preserve it. The same may be said for random mutations in the genome producing useful strings of amino acids (proteins) and their preservation. But this supposes that nature knows what is good for it.

Beyond the logical limits to the effectiveness of random evolutionary selections, nature provides a format for a rigorous statistical test of evolution. It rests on the phenomenon known as convergent evolution.

The emergence of organs similar in shape or function in animals of different species is referred to as convergent evolution. Such organs are designated as homologous if they arise by inheritance from common descent. The organs are analogous if the similarities satisfy the same need or function but were formed by independent evolutionary paths, rather than by a common ancestry.

Bats and birds, whales and dolphins all have members that find their way by sonar. Emitting a burst of chirps often inaudible to the human ear, the animal listens for the echo. Based on the delay and changes in frequency of the returning sound, a bat can chart its course between the blades of a whirring fan and a dolphin can find its dinner in muddy channel waters.

The emergence of similar sonar systems among these diverse species is intriguing but not necessarily surprising. These animals have many similarities. All resemble a more ancient form of land-dwelling quadruped. All fall within the same type, or phylum, of animals known as Chordata (animals with vertebrae). In brief, their genomes (i.e., the genetic information held on the DNA of their chromosomes) share a common background that started 530 million years ago. With such a long common ancestry it is to be expected that their genetic material contains many similar inherited genes with which to construct similar organs.

Animals of different phyla do not share this common ancestral history. Approximately 530 million years ago the basic anatomies of all currently existing animals, from sponges to vertebrates, appeared simultaneously.[3-7] Because all the phyla appeared suddenly and simultaneously, the different phyla do not share a common genetic history above the level of protozoans. They separated, in a bush (not a tree) of life, at their inception 530 million years ago. This complete separation allows us to calculate the statistical likelihood that analogous organs in different phyla evolved independently (i.e., via convergent evolution).

Perhaps the most spectacular example of convergent evolution is the eye. A variety of animals in different phyla have "developed" eyes that are morphologically very similar. These include the octopus (phylum Mollusca), insects and the now extinct trilobites (both are members of phylum Arthropoda), and almost all vertebrates, including humans (phylum Chordata).

THE EYE AND ITS EVOLUTION

Richard Dawkins is a reader in zoology at Oxford University. He avidly favors the thesis that random mutations are at the base of all evolution and is famous for defaming the "cave man" mentality of those who consider the possibility that a Guide may have imposed direction on evolution. Nonetheless, Dawkins acknowledges that "It is vanishingly improbable that exactly the same evolutionary pathway should ever be traveled twice."[8]

While convergent evolution does not necessarily follow the identical developmental path to reach a common end in two separate species, it is described as having produced two very similar organs (the eye for example) in two unrelated animals (the octopus and the human). How? Dawkins tells us that natural selection did the job: "It is all the more striking testimony to the power of natural selection . . . in which independent lines of evolution appear to have converged from very different starting points."[8]

The discoveries of molecular biology, especially those revealing the similarities of genes activating the development of the eye, have brought this assumed "independence" into serious question. Is it possible that the analogous development was indeed not independent?

Again in the words of Richard Dawkins, "Measuring the statistical improbability of a suggestion is the right way to go about assessing its believability."[8] Let's do just that.

Let us assume in the most optimistic case that all the genes needed for the complexity of eye development and function were neutrally present in some one-celled eyeless organism that was an ancestor common to all animal phyla. Mutations would then "merely" need to activate these preexisting, nonexpressed genes and the eye would form in the new animal. Although this stretches the credibility of the argument to its limit, it would certainly simplify the convergent process.[9]

The gene that controls the development of all eyes has been found to be the same in all phyla. This implies it originated in a putative pre-Cambrian ancestor. The puzzle arises as to why these ancestral eyeless forms of life would have harbored a gene that was eventually to direct eye development in higher organisms. If the gene was used for another purpose in the ancestor, the puzzle deepens when we consider that not only has the structure of the gene been retained in all visual systems of all the diverse phyla, but also its function in these diverse phyla is the same.

Let's take a very conservative guess at what it takes to make an eye. Assume there are just one thousand mutational steps that lead to an eye in a formerly light-sensitive but sightless species. If each mutation reflected a change of one nucleotide base on the DNA molecule, the one thousand mutations would represent a change in less than one millionth of the total number of nucleotides present in the genome.

A bundle of nerve fibers must extend from the brain (and we're not even considering the complex development required in the brain to make use of the electrical signals from each of the optic nerve fibers). The surface at which the light-sensitive cells are located must invaginate. This probably occurs gradually, not as single massive recession of that spot. The structures of the eye (cornea, transparent lens with muscles, iris with muscles, vitreous humor, pigments, retina with receptors connected to optical lobes in the brain, lateral inhibition, etc.) must all develop. A thousand steps is probably a minimum but in this analysis, if I err, I want to err "in favor of" evolution.

Some processes seem "forbidden." No eyes, for example, regulate light entering the globe by varying the transparency of the lens as is done with a variety of manufactured eyeglasses. Nature might have chosen this technique. It varies the transparency of human skin to regulate the amount of light penetrating the outer layers. Possible loss of fidelity in the image's color passing through a tinted lens may make the eye's method of light control preferable.

Because a similar eye has evolved in animals of two separate phyla, mammals and molluscs, we can investigate the statistical probability of independent random-point mutations in the DNA genetic package producing a similar eye twice. Had the eye appeared only in one phylum, that is, in animals having a common genetic history, the statistical analysis would have little or no significance.

The mutational changes must develop somewhat in order. Invagination is beneficial in the selection process only if the light-sensitive cells are already functioning and able to benefit from the added protection, added surface area, and directionality provided by the invagination.

Give nature a moderately free hand. Let's say at each step there are four options, four possible mutations. All DNA has four types of nucleotide bases. Combinations of three of them, a triplet of bases, code to form each of the twenty common amino acids used to build the proteins of all known life. We assume one of those four possible mutations will activate the next needed gene in the evolution of the convergent organ, here the eye.

TWO IMPLAUSIBLE MODELS FOR EVOLUTION

If the evolutionary model that we choose is such that each of the thousand steps in our hypothetical mutation must be in sequence and any erroneous or out-of-order mutation is fatal, the number of trials required in the process is 4^{1000}, or in the usual decimal notation 10^{600}. That is a one with six hundred zeros after it! And that is just to get the information of an eye to the brain. We didn't start the processing of that information by the brain.

But this model is too strict. All "erroneous" mutations, for example, may not be fatal. It is possible to envision a sequence in which the thousand steps can be accomplished with far fewer than the 10^{600} random trials.

With the statistics of probability, it is not the mathematics that is difficult. The difficulty is choosing a model that reasonably approximates the real world. If we take the most "optimistic" or forgiving set of assumptions in the thousand-mutation sequence, then the difficulty of achieving the desired organ fades to triviality. Allow mutations to occur in any order with no fatalities for incorrect mutations. Assume that the correct mutations are retained (that is, they are locked into the DNA and never mutated away) and allow all of the thousand potential sites which do not yet have the correct nucleotide base to mutate each generation.

Of course with such a high mutation rate, there is no chance that there would be no fatalities, but for this "forgiving" model we will assume there are none.

We want to calculate the likelihood (i.e., the probability) that a mutation will successfully cause the correct nucleotide to occupy the correct site on the DNA molecule and so produce the correct amino acid. The probability of success (p) is one minus the probability of failure (q), or:

$$p = 1 - q$$

In our extraordinarily optimistic model, each of the thousand sites acts independently. Only one of the four nucleotides is correct for each site. Therefore, three of the four are incorrect. Hence, in one trial

$$p = 1 - 3/4 = 1/4 = 0.25$$

There is a 25 percent probability that we will have a successful mutation with the first try. With multiple trials

$$p = 1 - q^r$$

where r is the number of trials.[10]

With this model, after a mere ten generations, there is a 94 percent probability that the goal will have been reached. With twenty generations, the probability of success is 99.7 percent. Quite a difference from our former 10^{600} trials.

Obviously, this overly forgiving model bears no resemblance to reality. Nonetheless, Dawkins uses a similar model to demonstrate the power of random mutations in the evolutionary process.[8] Dawkins took a random string of twenty-eight letters and then had a computer randomly change them to any one of twenty-seven variations—the twenty-six letters of the English alphabet plus one blank for spaces between words. Here we have twenty-seven variations per slot whereas in the previous example we had only four variations per slot.

In a mere forty-five generations, the letter string "mutated" into a previously selected verse from Shakespeare, "Methinks it is like a weasel."

With all letter slots "mutating" independently, each generation and all "correct" mutations (i.e., the correct letter in the correct position) locked in place, the probability equation shows that Dawkins's forty-five trials give an 80 percent certainty of successfully producing the chosen sentence.

Dawkins's success at forming his sentence proves only that his computer is working correctly! It proves nothing about evolution other than the reality that the model one chooses determines the results. Dawkins's model had a known goal and worked toward that goal, knowing which letters it wanted in each of the twenty-eight slots. It is an ideal demonstration of *directed* evolution. Yet he parades this model as a proof for the effectiveness of random mutations reaching a desired end. It is pure deception—I hope self-deception and not deliberate chicanery. With this model, in one hundred mutations we can produce not only a verse from Shakespeare, but all the works of Shakespeare and all the works of every other author that ever set pen to paper or finger to keyboard throughout all the history of mankind.

Somewhere between the two extremes lies the truth.

CHOOSING A REASONABLE MODEL FOR CONVERGENT EVOLUTION

I want to learn the statistical probability that nature will produce, independently by random mutations, two structures that are externally similar, although they may use different proteins in their construction. The parameters of this development are speculative. Here I choose a plausible though quite forgiving or lenient model as a means of speculating whether the process is within the realm of feasibility.

There will be no fatalities for wrong mutations. Each mutation will provide a 1 percent benefit (any larger benefit will by necessity demand that wrong mutations be fatal). Population size is 100,000 individuals.[11] In this model, mutations must be in sequence (e.g., no lens before we have a light-sensitive region). Each beneficial mutation will be permanently stored on the DNA. It is not allowed to mutate away.

The model, in a sense, "bends over backwards" to favor the theory of convergence and the appearance of the convergent organ.

The major uncertainty beyond estimating the number of mutations required to produce the convergent organ is the assumed rate at which the needed mutations occur. These mutations are significant only if they occur in sexually mature reproductive cells, that is, in gametes. A mutation in, for example, an individual's skin will not be passed on to that individual's progeny.

If we take data applicable to existing animals, reported mutation rates of gametes range from one mutation in ten matings to one per 100,000 matings.[12,13] Of course that mutation might not occur within the thousand bases we are seeking to change. If the entire genome, containing 3×10^9 base pairs in humans, is vying for the one mutation, then there is only approximately one chance in a million that the mutation will occur within the thousand bases of interest here.[14]

With a population of 100,000 individuals, even assuming the high reported mutation rate of one mutation in ten matings, there will be 5,000 base mutations per generation. If the genome of interest contains 10^8 base pairs, there will be only one mutation within the 1,000 bases of interest for each 100 generations. If the genome contains 10^9 base pairs as in humans, 1,000 generations are required on average. And this mutation may not be correct.

In the simplest model, each point of the thousand DNA sites under consideration has one of four base possibilities. Therefore in the thousand sites there are $4 \times 1,000$, or 4,000 possibilities, and all but one (i.e., 3,999) are incorrect. Again, the probability of success is one minus the probability of failure. The probability of success in the first of the thousand sites is

$$P = 1 - [(3999/4000)^n]^r$$

where P is the probability of success in r generations having n random mutations per generation.

In our example, $n = 0.01$, that is one mutation in the region of interest per one hundred generations. For these conditions of one mutation per ten matings and 100,000 matings per generation, a million generations are required to attain a 92 percent probability that the first of the thousand sites will be successfully filled. To achieve an 80 percent probability that all thousand slots will be filled with the mutants required to produce the convergent organ (in this example, the eye of the human matching, morphologically, the eye of the octopus), we require a trillion generations.

Generation times of protozoans are measured in days. The multicellular forms of life observed five hundred million years ago in the early Cambrian, if judged by their currently existing cousins, have generation times of weeks or months. These rates provide the opportunity for multiple generations each year. Even so, with the above model, the thousand mutations required to orchestrate already existing genes would require in excess of hundreds of millions of years. In contrast, the fossil record indicates that the Cambrian explosion occurred in five million years or less.[15]

The era of the Cambrian explosion represents a time approximately one hundred million years after the molecular oxygen concentration in the atmosphere rose to a level able to support large multicellular animals.[16] The increased availability of oxygen produced a tenfold improvement in the efficiency of energy extraction from consumed foods. This may have been one of the missing ingredients that had helped keep life single-celled for the previous three billion years. With the newly found energy, life could develop larger, more complex structures.

The potential for bigger, more complex life is the up side of oxygen. There is also a down side. Oxygen is a highly reactive element. If not controlled, it produces free radicals in a cell's cytoplasm which in turn can produce mutations in the DNA. The modern cell's mechanisms for repairing mutant DNA is an ongoing wonder of organization. Although millions of years had passed since reaching the high oxygen level, the early forms of multicellular life might not yet have developed the tolerance to the negative effects of the oxygen that we latecomers to the scene of life have. In addition, there is the possibility that the radiation-protective ozone layer may have not yet been fully established in the upper atmosphere. Hence the mutation rate in the early Cambrian may have been higher than that which is observed in modern gametes.

If we increase the assumed mutation rate a hundredfold, the population of 100,000 individuals will experience one mutation per generation within the one thousand DNA spaces of interest. This would be equivalent to ten mutations within the entire genome for each animal's mating. The convergent organ will now become dominant in the entire population after five to ten million generations. This includes the generations required for each mutation to spread throughout the population.

With the short generation times of these relatively simple animals, the convergent organ appears within the time frame presented by the fossil record. But keep in mind that to achieve this convergence we have boosted the rate of gamete mutations a hundredfold over the highest rates currently reported while maintaining the conditions that no mutations were fatal and all proper mutations were locked in—that is, they could not be lost by subsequent detrimental mutations. With a hundredfold increase in mutation rate, retaining these favorable mutations stretches plausibility beyond its limits.

More significantly, we assumed the genes were already present in an ancestor and merely needed to be activated by these mutations. If the genes themselves had to be formed by random reactions, the number of needed mutations increases by more than a hundredfold. Convergent evolution by random mutations of the DNA nucleotides becomes statistically so highly improbable as to be functionally impossible.

The implications of this have led researchers to report in the journal *Science,* "The concept that the eyes of invertebrates have evolved completely independently from the vertebrate eye has to be reexamined."[17]

An article in this most highly respected journal has asked for a reexamination of the process of evolution! The significance of this statement must not be lost. This genetic similarity is so extensive that it "strongly argues for a common developmental origin."[18] Convergent traits among animals of different phyla have challenged the very basis of evolutionary theory: the hypothesis that traits develop independently, initiated at the molecular level by random-point mutations.

Simply stated, the convergence observed in convergent evolution did not happen by independent, random reactions. It could not have happened by random reactions. It *must* have been preprogrammed.

This is merely an extension of the unlikelihood that life itself started by random independent reactions. Ilya Prigogine, a recipient of the Nobel prize in chemistry, wrote in *Physics Today:* "The probability that at ordinary temperatures a macroscopic number of molecules is assembled to give rise to the coordinated functions characterizing living organisms is vanishingly small. The idea of the spontaneous genesis of life in its present form is therefore improbable, even on the scale of billions of years."[19,20]

In these speculations we investigated changing one organ. But in the five-million-year transition from pre-Cambrian to Cambrian life, the basic anatomy of every animal alive today developed. Massive morphological changes were required in every part of the ancestral genome. Even more confounding to the traditional logic of evolution, there is no evidence of evolution within the five-million-year span of the Cambrian explosion.[21,22] Each of the animals in this era makes its first appearance fully developed. It was the sudden nature of this "evolutionary" development that led to Walcott's reinterrment of the Burgess Shale fossils. The idea of a hopeful monster, a massive and multifaceted evolutionary change, occurring in a single generation—or even in a few generations—simply does not stand up to the scrutiny of statistics. This was established in 1967 at the Wistar Institute symposium,[23] which brought together leading biologists and mathematicians in what turned out to be a futile attempt to find a mathematically reasonable basis for the assumption that random mutations are the driving force behind evolution. Unfortunately, each time the mathematics showed the statistical improbability of a given assumption, the response of the biologists was that the mathematics must be somehow flawed since evolution has occurred and occurred through random mutations.

The fossil record implies an exotic developmental occurrence at the Cambrian. These data are reported in the leading scientific journals. But how to explain the data is a mystery that becomes more mysterious with each new fossil. When Darwin wrote "natura non facit saltum" he was so very wrong—"gloriously and utterly wrong" (to quote Richard Dawkins, though admittedly out of context!). It seems that natura sola (only) facit saltum! Sudden appearances are the trademark of the fossil record.

Fossils reveal the events. Unfortunately, they do not disclose the processes by which those events occurred. It is true that hox genes have been discovered to control the location and development of entire organs. But the genes that actually form these organs must already be present in the genome. We are forced to revert to the idea of latent genes, waiting patiently to be cued by the environment for expression.

Hox genes themselves are a puzzle. They have been found to be highly conserved among diverse phyla. Are hox genes also preprogrammed? A nearly identical gene determines the position and orientation of the forelimb in humans and all vertebrates studied and also the position of the wing and its orientation in the Drosophilia fruit fly. The functioning is so similar that a portion of the vertebrate gene can be implanted in Drosophilia genome to induce wing development! The same is true for the gene controlling eye development in mice and in fruit flies.[24] If the fossil record truly records the history of evolution, then undoubtedly an exotic, in the sense of out-of-the-ordinary, mechanism is at work—be it preprogrammed information carried neutrally on the genome or Lamarckian-type environmental feedback influencing, even restructuring, the organization of the vast library of information carried on a genome. The answer to the question posed by the journal *Science,* "Did Darwin get it right?" is supplied by the journal itself: No, Darwin did not get it all right.[25] The same must be said for neo-Darwinian theory. Science may not be capable of adjudicating the issue of God's possible superintendence of nature, but it certainly has discovered that nature functions in a way that at time seems most unnatural.★

★The author acknowledges the contributions of Prof. A. Engelberg to calculations presented in this chapter.

CHAPTER 8

The Watchmaker *and the* Watch:

Concerning *the*

Statistical Probability of Chimps

and

Humans Evolving by Random

Mutations from a Common Ancestor

In crossing a heath suppose I pitched my foot against a stone and were asked how the stone came to be there. I might possibly answer that for anything I knew to the contrary, it had lain there for ever nor would it perhaps be very easy to show the absurdity of this answer. But suppose I found a watch upon the ground and it should be inquired how the watch happened to be in that place. I should hardly think of the answer which I had before given, that for anything I knew the watch might have always been there . . . The watch must have had a maker. There must have existed at some time and at some place or other an artificer who formed it . . . and designed its use.

Thus William Paley began his now classic treatise against evolution. It was published in 1802, fifty-seven years before Darwin's *Origin of Species*. Paley's approach, the argument that design implies a designer, is standard logic used in attempts to debunk the theory of random evolution. He titled this work "Natural Theology—or Evidences of Existence and Attributes of the Deity Collected from the Appearances of Nature." The title sounds like a rework of the opening of Psalm 19: "The heavens tell of God's glory and the firmament declares his handiwork." But Paley's argument does not go far enough and so falls short of the Psalm.

The argument of design assumes that since the complexity of a watch requires a designer, how much more so does the complexity of a human. Yet to an avowed evolutionist, Paley's logic is fatally flawed. The complexity of a watch obviously implies a watchmaker. Watches do not make themselves. But animals do make themselves and each animal is a bit different from its parents. Add up these differences over millennia, the evolutionary biologist will tell you, and you can get the complexity of a *Homo sapiens* out of a less complex ape, an ape out of a still less complex reptile, and a reptile out of a fish.

Nature strives toward complexity because complexity carries with it survivability through intelligent adaptability. The simplest form of life, bacteria, lacks this feature. Though as a group bacteria have been on Earth longer than any other form of life, as individuals they are not a success story. Their genetic DNA lacks any mechanism to retard or correct for mutant errors in the transcription. Every mutation produces either a new variety or a dead cell. Massive numbers of mutations occur, the overwhelming number of which are nonfit and perish. As a group they survive by sacrifice of the individual.

Let's examine the data and determine the statistical probability of complex human-like creatures evolving from apes. As the avidly pro-Darwinian evolutionist Richard Dawkins asserts, "measuring the statistical improbability of a suggestion is the right way to go about assessing its believability."[1]

The fossil most resembling a human skeleton is that of the Cro-Magnon *(Homo sapiens)*. First discovered in the Cro-Magnon cave of Dordogne, France, specimens have now been found across Europe, Asia, and the Americas. The similarity of Cro-Magnon fossils to the human skeleton includes hip bones for standing erect, facial bones, and the shape and size of the brain cavity. Our Cro-Magnon look-alike first appeared in the fossil record fifty to one hundred thousand years ago. By the time we reach ten thousand years before the present, there are enough fossils of Cro-Magnon to fill a modest-sized museum. These are dated by several independent methods and all methods give similar ages.

Either God put ready-made fossils in the ground for some

unknown reason or Cro-Magnon existed. A theologian's pretense that the data are the fabricated musing of demented godless scientists is counterproductive to all sides. Since the Bible defines a human as an animal with a *neshama*—the spiritual soul of humanity (Gen. 2:7)—there is no biblical problem with human-looking creatures predating Adam. As Talmudic and ancient commentaries point out, they were animals with human shapes but lacking the *neshama*.[2,3]

Cro-Magnon fossils overlap in age with the more primitive Neanderthal. Neanderthal resembles the usual comic strip pictures of cavemen: protruding brow, sloping forehead. Discovery of this overlap countered the theory that Neanderthal developed (evolved) into Cro-Magnon.[4,5]

Beyond the absence in the fossil record of an immediate ancestor to Cro-Magnon, there is the surprising genetic similarity among all humans. Considering the large number of our species and our wide distribution over the globe, there should be a wide genetic divergence among populations. That divergence is missing. Gorillas, for example, though geographically much more confined and very much fewer in number, show a genetic diversity far greater than humans. Our global similarity indicates that we humans all have a fairly recent, common ancestor.[6]

To study the origins of Cro-Magnon, I want to go further back in history than the tens of thousands of years of Cro-Magnon and Neanderthals. I want to return to the moment at which evolutionary biologists and paleontologists theorize a common ancestor diverged into two populations—one of which eventually became the chimpanzees and the other *Homo sapiens*. That divergence, we are told, occurred seven million years ago in Africa.[7,8]

For millions of years the topography of Africa posed no barrier to free movement. From the Indian Ocean on the east to the Atlantic on the west no major geologic formation blocked the potential for intermixing and interbreeding of the plethora of life on that very fertile continent. Eight million years ago this reality changed dramatically.

Shifting tectonic plates produced the Afro-Syrian rift, a ridge running south-north through Africa and continuing to Akaba/Eilat, Jericho, and Syria. The African portion of the ridge isolated the eastern third of the continent from the western two thirds. Animals in the eastern sector could no longer easily mix with and interbreed with their "relatives" living to the west. The tectonic shift had established genetic isolation, one of the traditional prerequisites for speciation.

The ridge was so steep that it changed the flow of air, and hence the weather patterns, across the continent. The western region retained the previous high levels of moisture. Existing arboreal vegetation continued to thrive. No new environmental challenges were presented to the animals in that region. Not so to the east. The sharp rise of the rift reversed previous patterns of air circulation. Now air stripped of moisture reached the region. Forests dwindled. The landscape changed to savanna grass with only occasional trees.

Evolutionary biologists believe that the new savanna environment favored creatures that walked upright. Over the next seven million years, we are told, a tree-dwelling primate adapted to this change by gradually developing (evolving) into an upright primate (the first hominids), and then to our Cro-Magnon look-alike.[7,8] Since the western two thirds of Africa maintained its abundance of forests, the putative common ancestor of chimps and Cro-Magnon living in west Africa may have changed only slightly during the seven million years following the formation of the rift. This would be reflected in today's chimpanzee.

A few hominid fossils have been found in western Africa. However, the large majority appear on the eastern third of the continent. If these western hominids indicate that east-west migration across the rift was possible, then the premise of genetic isolation fails and we have no clear indication as to what produced the change to upright bipedal walking.[9]

Most published opinions support, perhaps by necessity, the genetic isolation theory. It is a reasonable first assumption. Let's develop a reasonable model for the genetic changes that might allow a tree-dwelling ape to become an upright Cro-Magnon. We can then calculate the probability of it occurring in the seven million years available.

We read in the scientific literature that some primates have characteristics once thought reserved for humans. Bonobos have face-to-face sex, nurse for up to five years, become sexually mature at age thirteen to fourteen.[10] A chimp may groom itself while looking into a mirror (self-awareness perhaps). Genetically chimps are almost human. (Actually if a chimp's face is shaved or if I let my beard grow to an inch or so, I do see a resemblance.) There is an intense effort underway to study the entire human genome. Only a small fraction of the chimp genome has been studied. But if we take the known data as indicative of the entire genome, then humans differ *genetically* from the chimpanzee (genus *Pan,* hence the name) by 1 to 5 percent depending upon which sequence of amino acids (proteins) is being compared. Notice that nothing about the soul, the *neshama,* is being discussed here. That's left to spiritual, not scientific, scrutiny. Nobel laureate physicist Leon Lederman neatly summarized the dichotomy for me: "When people ask me to talk about spirituality, I tell them to see the people across the street [theologians]."

A 1 to 5 percent difference does not sound like much. Considering how we humans act at times, it may be an accurate accounting!

The genetic package of information carried on the chromosomes, known as the *genome,* is about the same size for all mammals. It consists of approximately three billion nucleotide base pairs per cell. These bases are the molecular points of information held within the double-helix-shaped strands of our chromosomal DNA. At each of those three billion points, one of four different bases resides. A set of three bases (a codon) codes for one of the twenty naturally occurring amino acids. A combination of several hundred linked amino acids folded into a specific three-dimensional structure produces one of the thousands of proteins found in life. Inherent properties of molecular binding and electrical charge of each individual amino acid change the linear 1-dimensional strings of nucleotide information into the 3-D structures of our proteins and hence our bodies. This change from 1-D to 3-D is by itself another of the beautiful subtle wonders which compose life.

Of the three billion base pairs, only about 3 percent appear to be expressed in the coding. The remaining 97 percent of the DNA seems to be inactive although it may be crucial to the functioning and regulation of the genome.[11]

Let's assume that to form a chimp and a Cro-Magnon from their putative common ancestor that lived seven million years ago, we need only change the active part of the genome. That entails 3 percent of the bases or (3% x 3 x 10^9 =) ninety million bases. Our DNA differs from chimps' by about 1 percent in the expressed region of our genome. Within these ninety million active bases, we'll assume most conservatively that there is a 1 percent difference between the putative ancestor of Cro-Magnon and the Cro-Magnon. (Recall that the human skeleton matches that of the Cro-Magnon.) That amounts to 1% x 90 x 10^6, or approximately one million differing bases.

That's the tally molecular biologists have set for us. The question is: can we expect to have one million point mutations in DNA during the seven million years available, and have those mutations retained and then have them passed on to become dominant traits in the entire local population or herd in that span of time?

Now comes the difficult part of probability: developing a reasonable model.

At first glance we might ask: which mutations are required to change a common ancestor similar to a chimp into a Cro-Magnon? But this approach has a basic flaw. In the eyes of the evolutionary biologist there are no "required" mutations. Required mutations imply a goal toward which natural selection (i.e., evolution) is heading, and we are told, time and again, that "natural selection lacks all intention."[12] The appearance of intent arises from hindsight, from looking at the product of evolution and marveling at the planning required to lead from an amoeba six hundred million years ago to, for example, the modern human.

If evolution indeed lacks intent, then we might deduce that if we humans did not evolve, one of a multitude of other forms of life, sufficiently intelligent and dexterous to meet the challenges of nature, would now be the top-of-the-line creature. Some being had to develop and we just happen to be it.

To summarize the neo-Darwinist position: We are the result of a series of random point mutations of DNA in gametes (mature sex cells) that united to form mutated zygotes that grew into mutated offspring. Nature's challenges then selected for or against the change expressed by the mutations. If selected for, then through many generations of interbreeding the beneficial mutation gradually became a dominant trait in the now improved herd. This random plus selective process was repeated a myriad of times for the needed million mutations until we reached the present. There is no fossil evidence for this, but we can work with it as a possible, conservative model.

If we take a different stance and claim that humans are the *only* top-of-the-line form capable of meeting the challenges of nature, the problem of producing us by the random plus selective natural process becomes insurmountable in the time available. Even if we assume mutation rates a million times higher than currently observed in gametes and even if none of the mutations is fatal (both being ridiculously expanding assumptions), we require more than one hundred million generations to reach the human. If we assume more normal mutation rates, the required time reaches hundreds of billions of generations! Any reader familiar with the equations of probability can make the needed calculations on the back of an envelope. There is no need to belabor the point here.

Let's just ask: in the seven million years since tectonic forces caused the change in African climate, can a set of mutant traits induced by a million mutations accumulate? We will allow the mutations to be random at the level of DNA and then selected for by nature's challenges. We assume no mutations are fatal and once a beneficial mutation occurs, it cannot be lost by subsequent detrimental mutations. Both assumptions are highly unlikely and forgiving, favoring the Adam-out-of-ape theory. If I err, I want to err on that side.

Maximum reported mutation rates in human body cells are about one per ten cell divisions. In gametes, mutation rates are much lower: from one per 10,000 to one per 100,000 generations.[13,14]

The lower rate for gametes is not surprising. Gametes are the source of the next generation. As expected, they have considerable biological machinery to guard against mutations and then to correct for any that get past the first line of defense. Gametes want to preserve the status quo. It is the mutation rate of gametes that concerns us here.

The approximate breeding population for herds of many large animals is ten thousand members.[15] Assume a yearly mating season in which all sexually mature members of the herd participate. At a gamete mutation rate even as high as one per ten matings, there will be five hundred mutations in the herd each year, with 3 percent of these, or fifteen, in the expressed region of the genome. At each

point on the chromosome, any one of four bases may reside. Therefore, on average only one in four of the fifteen mutations will lead
toward a positive improvement of the host animal. The mutations
must then spread through the entire herd by subsequent interbreeding of the mutant with its nonmutated relatives.

If the mutation is neutral, offering no immediate selective advantage, and it is not lost by subsequent mutations, it will spread through
the herd in approximately five thousand generations. If each mutation gives as much as a 1 percent advantage to the mutant over the
remainder of the herd, five hundred generations are required for
each new mutation to spread through the herd and become a dominant trait in the entire herd. The compound interest formula gives
this result. (You will notice that I am listing the assumptions and the
basis of my calculations. I want to avoid any "monkey" business in
this discussion of the power of evolution.)

If the common ancestor of Cro-Magnon and chimps was sexually
similar to modern-day chimps (a reasonable assumption), then it
attained sexual maturity at about seven years of age. The five hundred
generations of interbreeding thus represent over three thousand years.

All the individual traits need not have developed sequentially.
Changes such as hip structure, hair cover, cranial shape, and the like
may have occurred concurrently. To provide selective advantage and
allow interbreeding, the mutations must have occurred in logical
order within each trait, so that individual traits changed gradually
(such as the gradual alteration of bone shape from the original to the
new). Assume fifteen concurrently changing traits, each trait having
seventy thousand altered sites. As an added benefit, we will have the
mutations always fall within the narrow one million site band (out of
the total of ninety million active sites) that harbors the basis of
change. At each nucleotide location, one of four bases may be
selected, only one of which is correct. The probability, P, of filling
the first site correctly is one minus the probability of failure, q:

$$P = 1 - q = 1 - (279,999/280,000)^r$$

where r is the number of generations. The 280,000 represents the
70,000 sites of each trait, each site having one of four possible bases.

Some 500,000 generations are required for an 83 percent probability that the first of the 70,000 mutations will have occurred. For completion of the task, we require hundreds of millions of *generations.* The fossil record tells us it happened in seven million *years.*

Alternatively, if we allow the needed million mutations to occur in any order, but have the fifteen annual mutations distributed randomly over the ninety million bases, then there is a 17 percent chance that the mutation will occur at a potentially beneficial site and a one in four chance that the mutation will select for a beneficial base at that site. The probability of success in the first generation is:

$$P = 0.17 \times 0.25 = 0.0425$$

The probability of failure is:

$$q = 1 - P = 0.9575$$

The probability of success in subsequent generations is:

$$P = 1 - q^r = 1 - (0.9575)^r$$

Forty generations pass before there is an 82 percent probability that any one of the potentially useful mutations will have occurred. Nature requires a million changes or forty million generations, even neglecting the five hundred to five thousand generations required for each mutation to spread through the herd. The fossil record gives us seven million years.

It might appear that there's a simple solution to this problem. Perhaps we're starting with an artificially small population and a larger population will evolve more rapidly. Increasing the herd size to 100,000 members provides a tenfold increase in the number of mutations per generation. However, with a herd of 100,000, over a thousand generations are required for each mutation to spread through the herd. Increasing the size of the herd does not solve the problem of how to get the mutations in the needed time.

In an attempt to explain the amazingly rapid morphological

changes observed in the fossil record, theories of evolution today talk of genetic bottlenecks in which very small herds undergo very large mutational changes. The herd's small size allows a mutation to spread rapidly through it by breeding. But the smaller the herd, the fewer the number of mutations per generation and therefore the greater the number of generations required for a mutation to occur. The larger the herd the more mutations per season, but also, the more generations required for the mutation to spread through the herd. Confirmed evolutionists agree that you just cannot win if the classic concept of *randomness* at the point molecular level of DNA is the driving force behind the mutations.[16] The time is just not there.

Large groups of bases or even entire gene sequences in the genome must have been activated or inactivated simultaneously as units. Groups of genes used for one purpose must have changed functions simultaneously, with some genes waiting patiently and neutrally quiet in the genome for millennia until they received a genetic or environmental cue to join their redirected cousins.

We have discovered over the past few decades that the exquisitely tuned laws of nature produce these wondrous manipulations in the genome. DNA sequences known as Hox genes can move entire organic structures such as eyes or legs to new locations. On rare occasions small bits of DNA actually move from one chromosome to another thereby initiating sudden (punctuated?) changes in expression. The genetic package from the female acts in the embryo in ways different from the package received from the male, although both carry alleles of the identical genes. Some proteins are coded for by phantom, nonexistent genes, or if not phantom, then genes composed of spliced parts of other genes. These wonders of the genome have changed our understanding of the processes underlying "evolution,"[17-20] processes no longer viewed as arising simply from random events.

Our universe, tuned so accurately for the needs of intelligent life, indeed ticks to the beat of a very skillful Watchmaker.

CHAPTER 9

. . . .

The

Origin

of

Humankind

*[And the Eternal God] breathed into his nostrils the neshama of life
and the adam became a living soul. (Gen. 2:7)*

*"became a living soul": it may be the verse is saying that it [adam]
was completely living being and [by the neshama] it was changed into
another man. (Nahmanides, commentary on Gen. 2:7; written in the
year 1260)*

In 1948, an event occurred that was destined to change the course of
history and of humanity. In that year, the patriarch Abraham was
born in the Mesopotamian city of Ur. It is Abraham whom the Bible
credits with rediscovering the existence of a single, noncorporeal,
Eternal Creator. It was Abraham who set out to teach that truth to
all humankind.

The calendar which records Abraham's birth date as 1948 is biblical, a calendar that starts almost six thousand years ago with the creation of the soul of Adam.[1] The coincidence is intriguing: 1948 as
the birth of the father of the people of Israel, and, from another perspective, 1948 as the rebirth of the State of Israel.

Yet the Bible's dates provide more controversy than coincidence. According to the biblical calendar, Adam's birth occurred within the last six thousand years. Can this be true when museums are filled with human-looking fossils dating back fifty thousand years? A clue to the answer to this biblical conundrum may be found in events that occurred two thousand years after Adam, in the life and affairs of Abraham. Just as God chose Abraham only after Abraham chose God, so the Eternal's encounter with Adam may have followed Adam's recognition of that transcendent yet immanent omnipotence we refer to as the Eternal God. Adam may have been the first hominid with a divinely created human soul.

When we deal with origin of the universe, we are talking of events fifteen billion (Earth-perspective) years ago. It is remote and speculative. The start of life also occurred so long ago it might as well have been the dawn of time.

But with the onset of humanity, we're beginning to shape our family tree. It's a bit too close for comfort. The atheist often wants "Made by Monkeys" stamped right across our wonderfully high brows, while the theist often seeks to prove we are a direct line from the "dust from the ground" (Gen. 2:7). According to the book of Genesis and two-thousand-year-old traditional commentary thereon, the reality of our existence lies somewhere between these two extreme positions in this most contentious of issues in the controversy between science and religion.

It is possible to explain the hominid fossils that predate Adam as having been placed there as a test by the Creator. And that may be true. There certainly is no way of disproving this hypothesis. Here, I argue from a different tack: that the fossils record a true account of history, that they present no threat to Torah, and most important, that major ancient commentaries anticipated these discoveries.

The literal reading of Genesis does indeed relate that "the Eternal God formed mankind dust from the ground" (Gen. 2:7), and this *literal* reading may be literally correct. However, in line with ancient *mainstream* interpretation of the Hebrew text, it *may* be (and I emphasize the word may) that Adam had an ancestor. Hold on!

Don't tear this chapter out of the book just yet. In the previous chapter, I showed the improbability that random reactions could change an ape into a human in the time available. Now hear my argument for this possibility of our origins, keeping in mind that according to the Bible, what makes humans unique is our *neshama,* our soul (Gen. 2:7), not our bodies.

When I was first struggling with the questions of our origins, I steeled my courage to ask the renowned biblical scholar, Rabbi Aharon Lichtenstein, if it was possible that Adam had an ancestor. Not knowing what to expect, I skirted the issue for a few awkward minutes. When I finally presented the question, his matter-of-fact reply almost bowled me over: "The text of Genesis and the ancient commentaries of Nahmanides on that text certainly [certainly, mind you!] leave the door open for that interpretation."

I have since repeated my question to leading Bible scholars, Jewish and Christian, in the United States and Israel. Their answers have been the same, though they replied only on the condition that I not quote them by name. They warned me to beware of the resistance I would encounter in teaching this and cautioned that it requires a deeper understanding of science and of Bible than many persons have. I have confidence in human intelligence and so I teach it.

We may never know the full truth of our origins. No less an authority on evolution than Ernst Mayr, professor emeritus of zoology at Harvard University, former curator at the American Museum of Natural History, and avowed lifelong advocate of Darwinian evolution, has finally come to admit that the origin of our species is a "puzzle" (to use his word) that may never be solved. The link that leads directly to *Homo sapiens* is missing.[2,3] That should not be a surprise. Such direct "links" are not abundant in the fossil record.

I rest my case on the Talmud (ca. 500) and such biblical giants as Rashi (ca. 1050), Maimonides (ca. 1190), and Nahmanides (ca. 1260). Their commentaries were written centuries before paleontologists claimed to have discovered remains of human-like creatures (hominids) dating back fifty thousand years and more. What they wrote was what they derived directly from the Bible.

My approach to the topic is three pronged: (a) the paleontological record, (b) comparative anatomy, and (c) the theological record. With regard to the theological record, it must be stressed again that although the interpretations of these biblical commentators are strikingly similar to discoveries of modern paleontology, they were developed prior to and therefore independently of modern paleontology. They represent a meaning that lies within the biblical text.

THE FOSSIL RECORD

The uniqueness of our cosmic origins has been lauded by secular scientists. The big bang and the one-time inflationary force that immediately followed it are so precisely balanced toward the needs of life that the universe appears designed for life.

Following its creation, the universe raced outward in inertial expansion. In time, a massive nebula containing enough gases and dust to form a hundred billion stars flattened into a spiral-like galactic disk, 80,000 light years in diameter and 6,000 light years thick. We call that galaxy the Milky Way, named for the white swath it traces across the black of the night sky. It's our home in the universe, one out of a hundred billion galaxies that constitute our universe. Our closest large neighbor galaxy, M31 in the Andromeda, is two million light years distant. The spiral shape of the Milky Way was to reappear billions of years later in the curves of the nautilus seashell and the seeds on sunflowers.

Three billion years passed. The universe was now approximately ten billion years old. Thirty thousand light years from the Milky Way's center and just north of the central plane of the disk, primordial hydrogen that had mixed with the stardust of bygone supernovae agglomerated, drawn together by mutual gravitational attraction. Most of the mass of the agglomeration was drawn into its center. That infall of matter released enough energy to raise the central temperature to a million degrees Celsius, the threshold temperature for hydrogen fusion. The mass ignited and became a star, a giant furnace that squeezes hydrogen nuclei so tightly together they fuse, forming helium and releasing massive amounts of energy in the process. We call that star the Sun. Each second the Sun fuses 660 million tons of hydrogen into helium and energy.

The dust and gases not drawn into the Sun formed nine planets and a halo of comets. For the next billion years, the planet Earth, about 8 light minutes (or 150 million kilometers) distant from the Sun, was in the making. Most of that time it was molten, heated by the energy released as dust and rocks from space crashed onto its surface.

Gradually the surface cooled. Liquid water and dry land appeared 3.8 billion years ago. To the amazement of the scientific community, fossils demonstrate that life in the form of bacteria and algae appeared immediately after liquid water became available.

For the next three billion years, the only forms of life were single-celled organisms. Then some 530 million years ago, life exploded into the diversity we see reflected to this day. The fossil record reveals a staccato progression from the first simple forms of life to the complexity of today's biosphere. The pattern is not at all similar to the smooth fossil flow originally predicted. No less a scientist than Roger Penrose states: "To my way of thinking, there is still something mysterious about evolution." The statistical unlikelihood that it all happened by chance hints to mathematician Penrose of a flow "towards some future purpose."[4]

Some time between one and two million years ago, a creature having a cranial capacity a bit less than one liter appears to have walked upright. Data from this period are tentative at best, being based on only a few partial fossils.

By 150,000 years ago, Neanderthal had appeared. The fossil record for this period is more complete. Neanderthals were similar to modern humans in many morphological respects. A main difference was the shape of their skull. Although their cranial capacity was about that of modern humans, 1.4 liters, their jaw was more massive and their brow more sloped than ours. Stone tools found with the fossils appear to have been formed by deliberate chipping, a possible indication of ability to plan and to execute plans.

But something more extraordinary than the making of tools was underway. These creatures had started to bury their dead! We surmise this from the several complete or almost complete skeletal fossils found from this period, each placed on its side in the fetal position. Over 100,000 years ago, for reasons we can only guess, creatures decided that the dead of their kind deserved a fate other than being torn apart by the carnivores lurking outside their cave entrance.

By forty thousand years ago Neanderthal had disappeared, replaced by Cro-Magnon, though the two had coexisted for tens of thousands of years. The morphology of the new hominids was essentially what you and I are today. With the appearance of Cro-Magnon, we find a marked increase in motor skills. Finely shaped bone tools and trinkets now appear buried alongside the fossils. By ten thousand years ago, settlements of Cro-Magnon spread from France to Ukraine and across northern Canada. Stone lamps with places for wicks (evidence for use of oil or fat) and bone needles are found in the burial sites.

About ten thousand years ago the beginning of farming occurred in what is today central Israel and northern Syria.[5] A thousand years passed and we find reeds being woven into baskets. By eight thousand years ago pottery had been developed. All this, and yet, according to the dates of the Bible, no Adam.

The Bible talks about pottery, how it is used, its purity and contamination (cf. Lev. 11:33). But the Bible doesn't say a word about who invented pottery. It doesn't mention its invention because (I believe) that event predated Adam, and the Bible was well aware of this. The case with metalwork is very different. Genesis attributes the start of sophisticated forging of copper and brass to Tuval-Caine, the son of Lemech (Gen. 4:22). This was some seven hundred years after Adam, or about five thousand years ago.

Archaeological finds that predate the invention of writing five to six thousand years ago are referred to by the scientific community as prehistoric, because these finds are artifacts, and not deliberately recorded accounts of events. Cuneiform, the first form of writing, dates to the biblical time of Adam and Eve.[6]

Theology refers to the time of Adam as the beginning of humanity. Archaeology refers to the invention of writing as the beginning of history. Perhaps the two terms have something in common. In Chinese, the word for civilization is "the uplifting force of writing."[7]

The early fossil record is too complete and too well documented to pretend it is all a fantasy of some misguided paleontologists. We are not talking of one or two fossils such as the evidence of hominid life a million years ago. Fossil finds from six thousand to forty thousand years ago fill museums.

The dating of these fossils is based on several independent radioactive and thermoluminescent techniques. All give similar answers. The claim that the Flood of Noah's time (according to the Bible, 4,100 years ago) altered the fossils and therefore invalidated these dating techniques does not stand up to scrutiny. The creation scientists who cite this approach should reconsider. The early Bronze Age falls *after* the time of Adam but *before* the time of the Flood. The same dating techniques used for pre-Adam artifacts place the start of the Bronze Age at five thousand years before the present, just as does the biblical calendar. If the Flood had altered preflood finds such as those that predate Adam, it would have also altered the Bronze Age data. And it didn't. The pre-Adam data are valid.

But as we shall see, from a biblical perspective there is no need to deny the implications of the fossil data. In fact, they have come to confirm, not to question, the biblical account of our origins as elaborated in biblical commentaries of one and two thousand years ago.

THE RECORD OF COMPARATIVE ANATOMY

Persons who find pre-Adam hominids a theological threat may persist in questioning the accuracy by which these fossils are dated. There is, however, little room for doubting the data that come from comparative anatomy.

That the human body has features very similar to lower forms of life is well established. Portions of the proteins in our bodies are near replicas of those found in single-celled bacteria and oak trees, not to mention monkeys and mosquitoes. Some of these proteins are chains of over one hundred amino acids, all in the same sequence. Since there are twenty types of amino acids, the likelihood of chance reproducing the same hundred-long sequence is one in 20^{100} or, in the more usual ten-base system, one chance in 10^{130} (that number is a one followed by 130 zeros!). In other words the probability of this similarity having happened by unrelated chance is approximately zero.

The bones of our arms and hands have nearly identical counterparts in the forelimb and pectoral girdle (that is, the flipper) of the beaked whale, as well as the wing of the bat and the bird and the manus of the alligator, not even to mention the more obvious

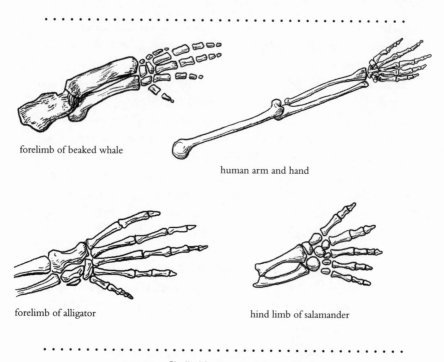

forelimb of beaked whale

human arm and hand

forelimb of alligator

hind limb of salamander

FIGURE 5: Similarities among tetrapod limbs

monkey (see Figure 5). This means that there are sites in our genetic material that produce the same organs as those of lower animals.

Even the one organ that would seem to make us different from every other animal, the brain, invokes an anatomy that is similar to lower forms of life. The human brain is layered. The lowest layer, the brain stem, controls automatic body functions such as breathing and heartbeat. Above this is the r-complex, r as in reptile. From here the reptile-like instincts of fight or flight and territoriality originate.

Most of us are familiar with that all too "human" emotion of territoriality. You have been searching for a parking spot for fifteen minutes with no luck and you are now ten minutes late for your appointment. Then you spot the one vacant space on the street. You pull up to parallel park but before you can start to back in, the car behind you pulls nose-first into "your" spot. Your first instinct is to get out of your car and throttle the other driver. That's territoriality at work. Does it mean that you have a reptile somewhere in your family tree?

Above the brain's r-complex is the limbic system, which gives rise to the emotions of caring for offspring and love. All mammals have it.

The uppermost layer is the cerebral cortex. From here originate such human facilities as language and analytical thought. From mice to monkeys to humans the structure of the cerebral cortex is the same, though it is by far the largest in humans. But the human cranial capacity (brain size) and shape is the same as the 1.4 liter brain found in forty-thousand-year-old Cro-Magnon fossils.

All of us in embryogenic and fetal form passed through stages where we had a yolk sac reminiscent of a fish egg, a tail, a three-chambered heart (similar to that of a reptile) that developed into the mammalian four-chambered heart, a reptilian double jaw joint, skin folds similar to gill slits, and a covering of hair (see Figure 6). These are things we'd expect to see in a zoo.

But with Adam, a change much more significant than a *quantitative* change from a medium brain to a super brain occurred. According to the Bible, and as we shall see according to archaeology, it isn't brain size or body hair that makes humans special. We are *qualitatively* different from all other life. What separates humans from all other forms of life is our soul of human spirituality, the *neshama* in Hebrew.

THE THEOLOGICAL RECORD: INTEGRATING BIBLE AND SCIENCE

More than 100,000 years ago, Neanderthals started burying their dead. At about the same time, Cro-Magnons also appeared. As fate would have it, the race between these two species for dominance appears to have been played out in the Middle East, the same region of the world where in later days the battle against paganism was to be staged. The race was to the Cro-Magnon. The Neanderthal disappear from the fossil record.

Until about forty thousand years ago, tools were limited to a few

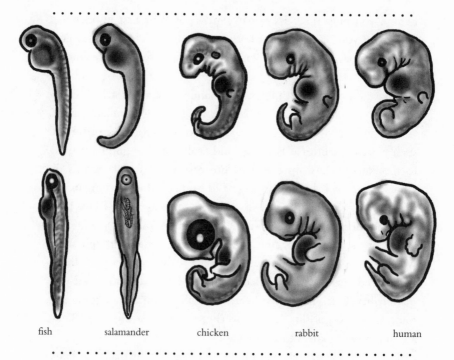

fish   salamander   chicken   rabbit   human

FIGURE 6: Several vertebrate embryos (The general form of these embryos is adapted from W. McGinnis and M. Kuziora, "The Molecular Architecture of Body Design," *Scientific American*, February 1994.)

The embryos in the first row are approximately four weeks old. Those in the second row are approximately six weeks old. Even at six weeks, with 15 percent of their term completed, the similarity between human and rabbit is obvious.

The development of a complex adult animal from a single fertilized egg is a wonder which happens so often that we have come to regard it as commonplace. When an abnormality occurs, we search for the reason. More logically we might marvel at the processes that produce the proper growth from egg to embryo to adult animal time after time.

Embryos develop by what appears to be a conservative process, in that they pass through developmental stages of other animals that might be seen as "lower" on an evolutionary bush. For example, fishlike yolk sacs appear and then disappear at one stage in the human embryo, having served no apparent function. The larval salamander has, at one stage, external filament-like gills. This progressive staging of the embryo led the German biologist Ernst Haeckel to suggest that ontogeny (the development of the individual embryo) repeats phylogeny (the history of the phylum). If we take this to mean that, at early stages, the human embryo appears very much like the *early* stages of a fish, then Haeckel's idea has merit. But the early stages of the human embryo never appear as an *adult* fish.

simple shapes. Then, in a cultural breakthrough, knives, sewing needles, statues, and cave paintings appeared. For the past ten thousand years, tools, food, and sculptures appear alongside the neatly buried hominids. Why did prehuman hominids bury their dead, and why with items that clearly required many hours of work to acquire?

CHOOSING GOD: THE ONTOGENY OF THE SOUL

The answer may lie in a quest that has a parallel in the biblical account of Abraham, two thousand years after Adam (see Figure 7). God told Abraham to leave his homeland and go to Canaan (Gen. 12:1). There is only the slightest hint given in the Torah as to why Abraham was chosen by God, but every Talmudic legend that elaborates on this event has the same message. Immediately after God commanded Abraham to leave, we are told, "Abraham took . . . the souls they [he and Sarah, his wife] had made in Haran and they went out to go to the land of Canaan" (Gen. 12:5).

Souls they had made? Isn't it God, not man, who is responsible for making souls? Biblical commentary tells us that these "made souls" were the persons whom Abraham and Sarah had already converted to the belief in one universal, noncorporeal God.[8] Even before God had spoken to Abraham, Abraham was actively spreading the word of his new realization. God was to choose Abraham only long after Abraham had chosen God.

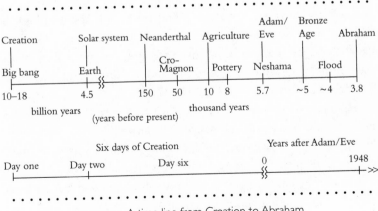

FIGURE 7: A time line from Creation to Abraham

Abraham grew up in Ur, a city located near the Euphrates river, 300 kilometers south-southeast of today's Baghdad. It was the heart of Mesopotamia and the fertile crescent. Abraham was born 1,948 years after Adam and Eve, or approximately 3,800 years ago. The extensive archaeological finds from the area indicate that in the time of Abraham an idolatrous religion was widespread.

Legend tells us that Abraham's father was an idol maker. But young Abraham had doubts that a stone figure made by a human could actually be a god. There must be something greater. First Abraham thought the stars must be the deities. Then the Moon rose, shining brighter than any star, and it seemed the Moon must be the ruler. But with sunrise, the Moon lost its splendor. And soon the Sun also set. At this point, the legends tell us, Abraham realized that there must be a supreme ruler, a Creator of the heavens and Earth who is not limited by the transience of material things. Abraham had discovered God.

In the Levant most people napped during the afternoon heat. The next afternoon, while Terah, Abraham's father, slept, Abraham smashed all except the largest of the idols in his father's workshop. He then placed a hammer at the base of the one remaining idol. When his father awoke, he was appalled, but Abraham explained that the big idol in a fit of jealousy had smashed all the others. There was the hammer right next to it as proof. "Don't be a fool," was his father's reply. "These are merely carved stones." "If they are only stone, father," countered Abraham, "then why do you worship them?" When Terah realized that his son was correct he took his family from Ur to start a new life: "And Terah took Abraham his son and Lot the son of Haran his [deceased] son's son, and Sarah his daughter-in-law, his son Abraham's wife and they went forth with them from Ur of the Chaldees to the land of Canaan; and they came to Haran and settled there" (Gen. 11:31).

Terah started for Canaan, the place where monotheism was to be nurtured, but he couldn't complete the journey. He could leave Ur, but he could not make the "Ur" leave him. When he found he could not abandon his idolatrous way of life, he settled in the city of Haran.

It was in Haran that God came to Abraham and told him to leave his father and continue on to Canaan (Gen. 12:1). Though Terah lived for another sixty years in Haran, Abraham made a complete break with the past. Now he needed a new environment in which to perfect his understanding of the world and mankind's place in it. Abraham, the one who had searched and made a choice, had become the chosen.

Just as Abraham was sent to Canaan to nurture his knowledge of God, Adam and Eve had also been set apart (in the Garden of Eden) to nurture their belief. Gradually, during the generations following Adam, the pure knowledge of a transcendental, noncorporeal God was replaced by a series of idolatrous intermediaries.[9] This transition has been confirmed by the abundance of idols found from that era in Mesopotamia and Egypt.

The biblical description of the upward flow of life during the six days of Genesis culminates in the making and creating of Adam: "And God said let us make *adam* in Our image . . ." (Gen. 1:26); "And God created the *adam* in His image . . ." (Gen. 1:27).

The making of mankind relates to the body of Adam. The Hebrew word *adam* has its root in the Hebrew *adamah,* meaning soil. The creation of Adam relates to the human soul, the *neshama.* Since legend tells us that Adam was created twenty years old, is it possible that an Adam-like being lived for those first nineteen years without the *neshama,* and then became human at twenty with the *neshama's* creation?

The Bible teaches that only at age twenty does one become divinely responsible for one's actions (Num. 1:3; 14:29; Deut. 1:39). It is our *neshama,* the spirit of the Eternal placed within us, not our physical attributes, that uniquely sets us apart from all other life, making us moral beings rather than amoral animals: "And the Eternal God formed the adam dust from the ground *(adamah)* and breathed into his nostrils the *neshama* of life and the adam became a living soul" (Gen. 2:7).

Let's study the Bible's wording more closely.

BIBLE LANGUAGE: "CREATION" VERSUS "MAKING"

As explained in chapter 3 on the age of the universe, the laws of relativity teach us that within the six days of Genesis there was sufficient time for the billions of years of cosmology and paleontology. In accord with the fossil record, the Bible teaches that plant life (Gen.

1:12) preceded aquatic life (Gen. 1:20) which preceded land animals (Gen. 1:24) which preceded humans (Gen. 1:27). But the Bible taught this three thousand years before the paleontologists discovered it to be true.

In Genesis 1:26, God talks of *making* mankind. In Genesis 1:27, the word *create* is used to describe the infusion of the soul of humanity. Since both these verbs characterize our human origins, there must be an essential difference in their import. To identify that difference, let's compare parallel uses of these verbs elsewhere in the Bible.

In Genesis 1:1 we are told: "In the beginning God created the heavens and the earth" (Gen. 1:1). In Exodus 31:17, we learn a significantly different account of our cosmic roots: "For six days the Eternal made the heavens and the earth" (Ex. 31:17). Is it "In the beginning" the heavens and the earth were *created,* or is it that it took six days for the heavens and the earth to be *made?* The objects of the two verbs, created and made, are identical in both verses: "the heavens and the earth." So we are talking of the same end product.

Creation, in biblical language, refers to the Eternal's introduction into the universe of something from nothing. It is an instantaneous act. Genesis 1:1 is teaching that in the beginning, in an instantaneous flash now known as the big bang, God created from absolute nothing the raw materials of the universe. Then, as Exodus 31:17 relates, this primordial stuff during the following six days was fashioned into the universe we know. Making, we see from the verse in Exodus, requires raw materials, and takes place over a period of time. As the Bible says: "For *six days* the Lord made the heavens and the earth [from the primordial matter created 'in the beginning']."

The universe was first created (Gen. 1:1) and then made (Ex. 31:17). That order was essential. Before the creation there was nothing with which to make.

For Adam, the order was reversed. The fact that Adam was first "made" (Gen. 1:26) and only later "created" (Gen. 1:27) informs us unequivocally that some amount of time passed during which Adam was fashioned. The *neshama* was implanted only after that vessel was complete. Whether that time was measured in microseconds or millions of earth years is not certain from the text. What is certain is that the making of Adam's body was not instantaneous and that its making preceded the introduction of the *neshama*. Making takes time. The ultimate change from the final form into a human was instantaneous, the creation of the *neshama*.

OUR ORIGINS: PART NATURAL AND PART SUPERNATURAL

The Bible explicitly states that the body of mankind was formed from the dust of the ground. The Hebrew word for man, *adam,* derives directly from the Hebrew word *adamah,* meaning ground or soil: "And the Lord God formed the *adam* dust from the *adamah* . . ." (Gen. 2:7).

The Bible also explicitly states that the bodies of animals were formed from the *same* material as Adam, the ground: "And the Lord God formed from the *adamah* all the animals . . ." (Gen. 2:19).

There is, however, a crucial difference in the original Hebrew between these two verses. The Hebrew word for formed, *ya-tsar,* when used for the forming of mankind, is spelled with two Hebrew letters *yud.* Although the structure and grammar are the same in verses 7 and 19, when used for the formation of the animals, *ya-tsar* is spelled with one *yud.* Every Torah scroll, whether from Yemen, Jerusalem, or Venice, California, is written this way.

Yud is the abbreviation of God's explicit name, best translated as the Eternal. As the ancient commentors, Rashi, Maimonides, and Nahmanides explain this verse, by doubling the *yud* for mankind, the Bible is telling us that although mankind and animals may share a common physical origin, there is an extra spiritual input in humanity. The *neshama,* the spiritual soul of humankind, is the factor distinguishing man from beast.

Hebrew has two words for soul, *nefesh* and *neshama.* They are represented by two divine creations mentioned in Genesis 1: "And God *created* . . . every animal . . ." (Gen. 1:21). All animals, humans included, share this first creation related to life. It signifies the infusion of the *nefesh,* the soul of animal life. A few verses later when the text tells of humans, there is a further creation, one in which lower animals did not share. "And God *created* the *adam* . . ." (Gen. 1:27). That creation marks the soul of humankind, the *neshama.*

This dual nature of humankind appears with the first mention of mankind: "And God said let us make adam . . ." (Gen. 1:26).

The "us" refers to the partnership, found in humankind, of the spiritual and material aspects of the world. Man has within him the animal as well as the godly.

AND THE ADAM BECAME *TO* A LIVING SOUL

The closing of Genesis 2:7 has a subtlety lost in the English. It is usually translated as: ". . . and [God] breathed into his nostrils the *neshama* of life and the *adam* became a living soul" (Gen. 2:7). The Hebrew text actually states: ". . . and the *adam* became *to* a living soul." Nahmanides, seven hundred years ago, wrote that the "to" (the Hebrew letter *lamed* prefixed to the word "soul" in the verse) is superfluous from a grammatical stance and so must be there to teach something. *Lamed,* he noted, indicates a change in form and may have been placed there to describe mankind as progressing through stages of mineral, plant, fish, and animal. Finally, upon receiving the *neshama,* that creature which had already been formed became a human. He concludes his extensive commentary on the implications of this *lamed* as: "Or it may be that the verse is stating that [prior to receiving the *neshama*] it was a completely living being and [by the *neshama*] it was transformed into *another* man."[10]

Another man! According to Nahmanides, who is the major kabalistic commentator on the Bible, the biblical text has told us that before the *neshama* there was something like a man that was not quite a human.

NONHUMAN CREATURES WITH A HUMAN MORPHOLOGY

That final form, before the creature became human, could it have looked human? There are ample ancient references to creatures, "beasts" in the wording of the Talmud, that lacked only the *neshama* to make them human. In one reference, this type of beast (rulers of the field is their formal name) is described as being so human-like in bodily appearance that its corpse may require the same respect as that given to a human corpse.[11] Recall that upon death, the *neshama* leaves the body. With the *neshama* now gone, there is no way of distinguishing a human corpse from the corpse of one of these beasts.

Maimonides, writing in the year 1190, refers to a Talmudic passage which discusses two verses from Genesis. The first tells us: "And Adam knew again his wife and she bore a son and called his name Seth" (Gen. 4:25). Four verses later we read: "And Adam lived one hundred and thirty years and was father to a son in his own likeness, after his own image and called his name Seth" (Gen. 5:3).

Here, as commonly found elsewhere in the Bible, a subtle fact is being implied by describing the same event from two slightly different perspectives.

Adam and Eve's first children were Cain and Abel. The Talmud deduces from these two verses that following the trauma of Cain murdering Abel, Adam and Eve separated. It was not until 130 years after Cain and Abel that "Adam knew again his wife [Eve]" (Gen. 4:25).

The Talmud asked why the Bible states "again" in reference to Adam's relations with Eve. Eve was Adam's wife, so obviously it was with her that he had relations. The "again" is superfluous and therefore it teaches something. The answer the Talmud supplies is that during those 130 years of separation, Adam had sexual relations with other beings (the nature of those beings is not clear). From these unions came children that "were not human in the true sense of the word. They had not the spirit of God . . . It is acknowledged that a being who does not possess this spirit is not human but *a mere animal in human shape and form* [!]. Yet such a creature has the power of causing harm and injury, a power which does not belong to other creatures. For those gifts of *intelligence and judgment* with which he has been endowed for the purpose of acquiring perfection, . . . are used by him for wicked and mischievous ends."[12-14]

Here we have ancient accepted sources that describe animals with human shape, form, intelligence, and judgment. Suddenly cave paintings that predate Adam by twenty thousand years and ten-thousand-year-old inception of agriculture become understandable. These less-than-human creatures had human-like skills. What they lacked was human spirituality.

With whom did Adam have relations? Eve was the only other human woman there. The ancient commentaries imply that there were other mates available to Adam though they were not human. The fossil record might refer to them as Cro-Magnon creatures.

Having been deprived of Eve's spiritual contribution, the offspring of such relations would indeed be less than human.

We now know that the woman contributes essentially all of the cell's vitally important mitochondria. A subtlety in Hebrew grammar indicates that the woman also contributes a large part of the *neshama* as well.

The Hebrew language shows gender more clearly than does English. The two Hebrew names for soul, *nefesh* and *neshama,* as well as the term for the indwelling of the spirit of the Eternal (the *Shehina,* in Hebrew) and the three types of cognition by which God created the universe (Prov. 3:19, 20) are all feminine words. Perhaps the essence of these Godly manifestations is more strongly evident in women than in men. Lacking the input of Eve, these children of Adam would be inferior in spirit though similar in body. Seth, having been raised by Adam *and* Eve, had their likeness and their spirituality.

In referring to the generations following Adam and Eve, the Bible states that: "The Nephilim were in the land in those days and also afterwards . . ." (Gen. 6:4). The word Nephilim comes from the Hebrew root for fallen or inferior.[15] Adam, having a *neshama,* would find the Cro-Magnon inferior in spirit if not in body.

There is a Talmudic legend that lends credence to this account. Eve, upon eating the forbidden fruit from the tree of the knowledge of good and evil, realized her fatal error. She was destined to die. But perhaps Adam would live forever. In that case, the Talmud has Eve reason, Adam might take another wife. At the very thought of her beloved Adam having another wife, Eve gave him also to eat of the fruit so that he too would be mortal.[16]

That legend might be no more than an interesting bit of psychology except for one aspect: the possibility of Adam's taking another wife. Adam and Eve were the only humans at the time! So how did another woman enter the picture? Is the Talmud proposing that there were other females, though not human, present?

Today a hornet's nest of religious controversy hums about claims for the existence of pre-Adam creatures similar to humans in shape and intellect but less than human in spirit. However, when the ancient commentaries were being written paleontology did not exist. No one was digging up bones and claiming to discover relics of intelligent prehuman creatures.

No incentive existed to prompt commentators one and two thousand years ago to speculate about prehuman hominids. The pre-Adam life they revealed was learned directly from the special wording of the Bible. Science has confirmed the Bible's description of life as starting immediately on the cool Earth, of simple life forms developing to the complexity of the modern biosphere. Science has also confirmed the biblical assertion that less-than-human creatures with human-like bodies and brains existed before Adam.

WRITING HUMANITY IN THE RECORD

The Bible describes a qualitative leap, a creation, at Adam that occurred almost six thousand years ago. Archaeology has provided us with an impressive record of that change. Though the soul leaves no material remains, the effects of the spirituality brought by the *neshama* are written loud and clear in the remnants of ancient Mesopotamia.

Every museum I have visited sets the division between prehistory and history at the invention of writing—not writing with an alphabet (or better said, with an aleph-bet since the Hebrew aleph-bet predates the Greek alphabet by centuries[17]), but with pictographs that over several centuries developed into cuneiform's six hundred stylized symbols. That invention occurred in Mesopotamia, which was to be Abraham's home two thousand years later. Archaeologists date the first writing at five to six thousand years ago, the exact period that the Bible tells us the soul of Adam, the *neshama,* was created.

The parallel between Adam and writing is no mere coincidence. The stimulus that led to its invention is revealed by the early writing itself, all of which relates to keeping records of commercial transactions. The timing corresponds with the rise of big cities, one of the first being Uruk on the banks of the Euphrates River in Mesopotamia, some 100 kilometers northwest of Ur.

With the advent of large cities, the need for civil administration and the division of labor which inherently accompanies any move from farm to city made the invention of writing inevitable. In the Mesopotamian rooms of the British Museum there is a graphic description of this event: "The earliest cities: large towns evolved before 5500 years [ago] as local centers but about this time there was a change in scale. . . . Much the most important innovation of the period, however, was the invention of writing. . . . The stimulus for the emergence and development of writing was the need to record economic transactions." The start of writing relates to the need for bartering, which relates to the formation of large cities. The question that remains is: why did large cities form at that time?

With the invention of agriculture, the land could support large populations. It is, however, unlikely that a population explosion was the sole cause of the start of large cities. Hominids had invented agriculture almost four thousand years before the creation of Adam and for those thousands of years they did not live in large cities. A key element that would facilitate the transition from village to city was missing. I propose that the missing factor was the *neshama*.

The *neshama* instilled in humans knowledge of a spirituality that transcends the individual. It provides the potential for goals other than the physical, allowing for social relationships, the intent of which exceeds the desire for survival and physical gratification.

Many animals choose and want and act on those drives. These are what Professor Harry Frankfurt of Yale and Princeton universities describes as first-order desires. But humans, because of what may be their unique ability for self-examination and introspection, have the capacity for second-order desire, the desire to want to be different. Humans are capable of wanting to improve their preferences and goals. These are not simple, monolithic wants. Rather they are based on a fabric of life experiences and values that together draw for the individual the image of a person that individual *wants* to be.[18] The *neshama*, having provided the knowledge of a transcendental excellence, allows us to identify with that goal even though we have not yet achieved it.

For example, a miserly person may wish she were charitable and could give joyfully rather than grudgingly. Within each of us there are two identities: the me I am and the me toward which I strive. The *neshama* is a tool for personal growth.

The Onkelos translation of the Bible (ca. 150) from Hebrew to Aramaic captures this spiritual rise of humanity. The descriptive essence of animal life, *nefesh haiyah* in Hebrew, appears three times in the opening chapters of Genesis. First it relates to the inception of animal life on Earth, the aquatic life on day five (Gen. 1:20), then to the first land animals (the start of day six; Gen. 1:24), and finally to humans near the close of day six.

In each of the first two instances, Onkelos translates *nephesh haiyah* as a living animal, the literal meaning of the Hebrew. For mankind, the Hebrew text has a slight variation, saying that the *adam* became "to a *nefesh haiyah*" (Gen. 2:7). Based on the addition of the word "to," Onkelos translates *nefesh haiyah* as "communicating spirit": "And the Eternal God formed the adam dust from the ground and breathed into his nostrils the *neshama* of life and the *adam* became a communicating spirit" (Gen. 2:7). The vital element that sets humans apart from all other animals is our immanent spirituality and our ability to share that spirituality with others.

Archaeologists can never discover the fossil remains of *neshama*. It is totally spiritual. That notwithstanding, archaeological evidence has confirmed our biblical heritage.

As it is written: "The truth shall spring from the Earth" (Ps. 85:12).

CHAPTER 10

The

Science

of

Free Will

"To be or not to be." Choice is the fundamental and continuous characteristic of our lives. Within the possible range for our choices we wrestle with Shakespeare's most basic quandary: "Whether 'tis nobler in the mind to suffer the slings and arrows of outrageous fortune, or to take arms against a sea of troubles, and by opposing end them? To die: to sleep; No more; and by a sleep to say we end the heartache and the thousand natural shocks that flesh is heir to, 'tis a consummation devoutly to be wished. To die, to sleep" (Hamlet 3:1).

The Bible, three thousand years before Shakespeare, defined our alternatives in the same terms: " I set before you life and death, the blessing and the curse, therefore choose life that both you and your children may live" (Deut. 30:19). The contest is between the body's desire for the ease of eternal sleep and the soul's striving for worldly and spiritual excellence.

Before we enter an examination of why choice is between life and death and not, as we might superficially suppose, between good and evil, or how we might choose life over death, we first should know the limits of our free will. We live in a world governed by the laws of nature. Perhaps these laws predetermine our future path. Our genetic code, the DNA of our cells, might control our emotions and our actions. And beyond the constraints which the physical world may place on freedom of choice, theology seems to obviate the potential of free will. We are told that our Creator knows the future. If our future is foretold, our freedom may be an illusion.

The paradox can be divided into three aspects:

1. Is the world deterministic? Do the laws of nature, cause and effect, determine all future events?
2. Are we so completely programmed by our genetic code, our DNA, that our future desires are beyond our control?
3. If God knows the future, does it matter that physics or biology might allow us choice?

If we get beyond these questions, we can deal with the Hamlet within us and how we decide whether "to be or not to be."

THE PHYSICS OF FREE WILL: THE UNPREDICTABLE FUTURE

At the outset of this investigation we must lay to rest the once-popular concept that the future is predictable.

For 150 years, classical philosophers had a love affair with the theory of determinism. Pierre Simon de Laplace (1749–1827) used arguments of the purest simplicity to demonstrate that the future is predictable and therefore predetermined. He based his thesis on a most fundamental law of nature: that of cause and effect. A given cause always produces the same effect. This seemed so obviously true that the surprise was not that the future was predetermined but that no one had discovered this "determinism" before.

On a macro-scale we observe predetermined fate continually. If we release a ball on a smooth slope, it rolls down the slope and never up. The causal force is gravity; the effect is the downward motion of the ball. If we leave a cup of hot tea unattended in a room that is at a normal temperature, the tea cools, always. It never heats up. The cause is a law of nature found in thermodynamics which directs systems always to evolve from states of high energy (hot tea) to lower energy (tepid tea).

Laplace argued that what we take for granted on the macro-scale actually functions everywhere. Causality, the ubiquitous law of cause and effect, directs the flow of all events in all types of environments. Living systems are not different from inorganic systems. The chemical reactions within our bodies, be they related to the digestion of our food or a nerve-muscle reaction to a mosquito bite, are all definable by the same laws of physics and chemistry which govern not only Earth, but also the entire universe.

If this is the case, there is no room for free choice. The chain of events that led a person to choose a particular fork in the road may be long and complex, but each link in that chain of decisions was formed by the events that preceded it and each of those preceding events was governed by unchanging, all-powerful laws of nature. Extend this logic to the entire universe and it is but a small step to see that all events everywhere are predetermined by the events that precede them.

In 1927 a revolutionary concept pulled the rug from beneath the logic upon which Laplace had built his theory of determinism. In that year, Werner Karl Heisenberg published his principle of indeterminacy, the uncertainty principle. This defined a limit to the precision by which the position and momentum (mass times velocity) of any particle could be measured. The more closely one determined the momentum of an object, the less precisely one could measure the object's position. The exact value of both can never be measured. It is not a matter of waiting for a better "ruler" or a better "speedometer." The best of these instruments, even in the world of theoretical fantasy, will always affect the condition of that which is being measured.

For the first time, the scientific community admitted that there was a limit to scientific knowledge. Not being able to know the present exactly obviously meant that the future could not be foretold.

Heisenberg's theory was rapidly developed by such giants of physics as Wolfgang Pauli, Max Born, and particularly Niels Bohr into what became known as the Copenhagen interpretation of the uncertainty principle. In essence, they saw the uncertainty principle as leading to a realization that there is no one specific reality in the physical world. All the possibilities for existence that fall within the uncertainty of the measurement might actually exist, and only when we make an observation at one specific point do the other possibilities vanish.

According to this theory, which forms much of the basis for quantum mechanics, objects in the universe have extended, fuzzy boundaries. Being fuzzy, there are no exact edges to measure. Recent experimental data indicate that the fuzziness is real. In this research, when the fuzzy extensions of several particles overlapped, the particles actually merged into a single large entity.[1] This confirmation of QM theory is not surprising. Quantum mechanics has a sixty-year track record of predicting correctly the outcome of experiments.

We are reluctantly being forced to abandon our concept of a world built of classical subatomic particles (protons, neutrons, electrons, etc.)—classical in the sense of entities having distinct edges like microscopic ping-pong balls. Subatomic particles are better understood as being quantum objects, infinitely extended through all of space by some currently unknown and immeasurable phenomenon that possibly resides in a dimension outside of time and space.

(If you are content to accept the claims of QM, and wish to avoid the complexities of its proof which I am about to present, I suggest that you skip a few pages forward to the section on The Biology of Free Will.)

THE FALL OF ABSOLUTE CAUSALITY

For free will to exist, causality—the thesis that identical causes produce identical effects—must not be universally true. There must be some slack in the laws of nature. A classic experiment which demonstrates this intriguing (and according to Schrodinger "worrisome") quality of nature, that causality is *not* universally true, is the double slit experiment. It rests on the observed fact that electromagnetic radiation (microwaves, radio waves, light rays, X rays, gamma rays) and the subatomic matter (electrons, protons, neutrons, etc.) of which we and all the universe are composed exhibit properties that can only be described as arising simultaneously from waves (fields of energy) and also from particles (discrete entities). This wave–particle duality is a paradox of nature with which Einstein grappled unsuccessfully during the final decades of his life. How can something be both a wave and a particle?

In an attempt to investigate this phenomenon, we need to learn a bit about waves. A basic concept taught in elementary oceanography is the behavior of ocean waves. A wind blowing steadily over a large open expanse of water adds energy to the water and as a result produces a series of waves moving in parallel. Upon encountering an artificial harbor made of two vertical walls with a central passage for boats, the waves will proceed directly through the passage, provided that the opening of the passage is wider than two wavelengths (a wavelength is the distance between two consecutive wave crests or two consecutive wave troughs).

If, however, the passage in the breakwater is equal to or smaller than the wavelengths, then the waves will "feel" the edges of the narrow harbor opening as they pass through. Instead of moving straight ahead, they will bend to produce wave fronts that propagate into the harbor in semicircles. The line of the breakwater forms their diameters, with the center of the breakwater opening at the center of the diameter. This phenomenon is known as diffraction and is a basic characteristic of all waves. (See Fig. 8.)

If there are *two* narrow entrances to the harbor, then as the semicircular waves spread from each opening, at some places crests from the two entrances will coincide. There they will combine to produce a relatively high wave. At other locations, about half a wavelength away from these peaks, a crest and a trough will coincide, canceling each other and producing a spot of calm water.

FIGURE 8: Water waves passing through two sizes of harbor openings. *(Left)* Harbor opening wider than distance between waves (wavelength). *(Right)* Harbor opening narrower than distance between waves (wavelength).

Thomas Young, in 1803, demonstrated that light behaved in the same manner as these harbor waves. It is the kind of experiment you can repeat on a sunny day with a razor and some aluminum foil.

Young allowed sunlight to shine on an opaque plate in which two slits had been cut. A meter or so beyond this plate was a screen onto which shone the light passing through the slits. Slits that were wider than the wavelength of light produced an image on the screen that had the shape of the two slits. However, as the slits were made more narrow, the light diffracted and the shape of the projected image changed. Since wavelengths of visible light are between 0.4 and 0.7 thousandths of a millimeter (rather than the meter or so wavelength of water waves), the opening must be accordingly narrow for light to produce a diffraction pattern.

When only one of the narrow slits was open, a band of light with fuzzy edges appeared on the screen (Figure 9, top). This could be the

FIGURE 9: *(top)* Pattern of light on a screen after passing through an opaque plate with one slit open. *(Bottom)* Pattern of light on a screen after passing through an opaque plate with two slits open

result of diffraction (if light indeed propagated as a wave) or it might be that light was a particle but the narrowness of the slit somehow affected the particles as they squeezed by the slit edges.

When both narrow slits were opened, the wavelike nature of light was proven. Instead of having a pattern on the screen that was the sum of two fuzzy bands of light, there appeared a series of alternating light and dark bands (Figure 9, bottom). Young correctly interpreted these as being the result of light waves passing through the adjacent slits, diffracting, and then adding or canceling as the diffracted waves overlapped on the screen. Just as in the harbor, when two crests met, a peak was formed (here a bright band). When a crest and a trough met, a calm spot appeared (here a dark, lightless band).

Since only waves produce such banded diffraction patterns, Young had demonstrated that the nature of light was that of a wave. The proof was conclusive. Or was it?

THE DISCOVERY OF WAVE-PARTICLE DUALITY

In 1905, a century after Young's conclusive demonstration, Einstein came along and upset the cart. In that year he published the results of experiments that demonstrated what has become known as the photoelectric effect. In 1921 Einstein was awarded the Nobel prize for this work.

Light, shining on certain metals, knocks free a stream of electrons which if collected produces an electric current. This is exactly the effect used in the photoelectric switch that keeps an elevator door from closing on your leg. Einstein demonstrated that the rate at which electrons are emitted from metal is related not only to the intensity of the light beam but also to the "color" of the light. If the color of the light was kept constant but the intensity of the light changed, the energy of each emitted electron remained constant but the rate at which the electrons were emitted from the metal changed. Holding the light intensity constant but changing the color of the light produced a stream of electrons constant in number but with energies that varied with the color of the light, red producing the lowest energy electrons and blue producing the highest. With some metals an intense red light did not liberate any electrons, while even a dim violet light liberated many.

Einstein interpreted these results as a demonstration that the light was arriving not in waves but in packets of energy (to be called photons). A high-intensity light beam had more photons than a dim light beam. However, the energy of *each* photon had nothing to do with the light intensity. Photon energy was related only to the color of the light. Red (the longest wavelength and lowest frequency) is the lowest energy photon in visible light. Blue or violet (the shortest wavelength and highest frequency) is the most energetic visible photon. That is why even a weak violet light might knock electrons free while a powerful red light could not liberate any. The individual photons are what hit the electrons. Each individual photon must have at least a required threshold or minimum energy to liberate an electron.

If light is composed of particles, called photons, how did Thomas Young observe that light passing through narrow slits in a plate produced an interference pattern, a pattern associated only with waves and never with particles? A particle is a discrete entity. It may be able to bounce off another particle, but how can it *cancel* another particle (to produce the observed dark bands)? The photons do just that when the "crest" of one passes through the "trough" of another. Since when do crests and troughs relate to particles? They are characteristics of waves.

Ninety years after its discovery, this discrepancy has yet to be resolved. And to complicate matters even further, the double slit experiment has been performed using classical particles such as electrons and even such relatively massive items as atoms, as well as photons. All behave as waves and as particles, notwithstanding the fact that this is patently impossible! In theory all matter, even you, has this same duality.

Bohr pointed out that this paradox of duality has strong implications relative to our knowledge of the subatomic world. If we measure an entity in a way that assumes it is a wave we find a wave. If we measure the same entity and assume it is a particle we find a particle. We see the world as we assume it exists.

CAUSALITY DEFEATED

The experiment about to be described can be performed with any of the wave-plus-particles of the universe. Electromagnetic radiation (microwaves, radio waves, light, X rays, gamma rays), electrons, protons, and possibly even hammers or elephants are suitable (though it is considerably easier to perform it with small particles). Here we use a maser, a gun that can fire one atom at a time.

Suppose the gun fires an atom toward a plate that has the traditional two slits in it. Beyond the plate is a screen with photographic emulsion. If an atom hits the plate, it is stopped and seen no more. If it happens to pass through one of the slits, it continues and strikes the emulsion, producing a spot on the film.

With only one slit open, we continue to fire the gun, one atom at a time. After a large number of spots accumulate on the emulsion, we notice that they have produced the expected fuzzy diffraction pattern described in Figure 9, top, above. Now we close the first slit, and open the second slit. Repeating the firing produces the same pattern (Figure 9, top), but offset by the distance that separates the two slits. The atoms are producing the diffraction pattern characteristic of waves passing through a narrow harbor opening.

Now we open both slits and again fire one atom at a time. The individual atoms no longer land randomly within the diffraction pattern (Figure 9, top). Instead, they fall only within the specific "allowed" regions where the light bands of the interference pattern appeared and never in the dark band regions (Figure 9, bottom). Seems reasonable, doesn't it?

But wait! This cannot be. We fire a single atom at a time. There is no other atom, be it wave or particle, with which to interfere and cancel. Yet the interference pattern occurs and the dark bands appear. A single particle can only go through one of the slits. We already noted that atoms going through the single slit fall everywhere within the diffraction pattern with none of the alternating light and dark bands that result from the interference of waves at the screen. Although we have opened both slits, we are still firing only one atom at a time. It must travel to only one of the two slits and go through that slit. If the other slit is closed it lands anywhere within the diffraction pattern. If the other slit is open, it never lands in the dark (forbidden) regions originally seen in the interference pattern which developed when we had the two slits open.

The atom is a single entity, with a fixed locality. In its passage through one slit, why should opening or closing the other slit have any effect upon its passage? How can it "know" if the second slit is open or closed? But it does know! Somehow it is aware of its environment.

The identical results are obtained when firing single photons. Photons are particles of light that travel at the speed of light, the maximum speed attainable in our universe. Even if the photon is infinitely extended, in the time it travels from the photon gun to the open slit it cannot have "felt" the second slit, checked to see if that second slit was open or closed, communicated that information to the portion passing through the first slit and then decided where on the screen it was permitted to land and where it was forbidden. There was no time for the feeler to make the round trip.

This is bizarre. With only one slit open, the particle could land anywhere within the fuzzy region marked on the screen (Figure 9, top). With two slits open, this is no longer true (Figure 9, bottom). There are forbidden regions, the dark bands. Ah, you might say, but that is because the wave property of the particle has canceled out those particles that might have landed in the dark region. Each wave crest, you claim, coincided with the trough of a second wave coming through the other slit and that produced the dark band.

When we were shooting millions of photons or millions of other particles a second, then we could imagine that many particles were passing through each slit essentially simultaneously and therefore many were arriving at the screen simultaneously. A 100-watt light bulb emits about 200 million million million (2×10^{20}) photons each second. They were able to arrive simultaneously and cancel or add their wavelengths.

In our present experiment this is not the case. We are shooting one particle at a time. It alone flies toward the screen. It passes through only one of the slits since it is only one particle. For some reason these single particles are unable to force their way onto the "forbidden" locations.

Two questions arise. If it is interference (adding or subtracting wave crests and wave troughs) that causes the alternating light-dark pattern, with what is the particle interfering? Only one particle is fired at a time, and we wait an hour between shots. (It is a very long experiment, but it gets the point across!) Only one particle at a time passes through the slit and arrives at the emulsion, yet that particle is somehow *interfering with itself!*

And equally puzzling: how does the particle know if the second slit is open or not? *No one knows* how or why the particle knows. But it knows.

There is a modification of the double slit experiment that could drive a particle physicist to become a carpenter or a biologist or anything other than a particle physicist. Put a particle detector near one of the slits and leave the other slit entirely unchanged. Continue to shoot one particle at a time. Now we can tell through which slit the particle traveled. If detected, it passed through the monitored slit. If not detected, it passed through the other slit.

The first thing we notice is that a whole particle always arrives at the monitored slit. A part of a particle never arrives. This means that the particle is not splitting into two half particles during its flight and trying for both slits. It goes discretely to one or the other slit.

With the detector in place, something very annoying happens. The pattern that accumulates on the screen as the experiment proceeds is the sum of two fuzzy patterns as if the first slit was open and the second closed, and then the second slit was open and the first closed. (Figure 9, top, plus this figure slightly offset). There are no dark "forbidden" regions even though both slits are now open and the banded interference pattern should appear.

One could argue that the detector alters the course of the particle passing through the monitored slit. Perhaps. But if it is true that the detector affects the monitored slit, what effect can the detector have on the other unmonitored slit? Particles passing through the second slit (the one with no detector) should follow the usual two-slit pattern. But they don't. They too somehow know about the detector at the *other* slit.

Not only do the particles know if the second slit is open, they know if someone is looking over their shoulders with a detector!

These experiments marked the end of the line for causality. In classical physics, causality requires that if the initial conditions are identical, the outcome must be identical. In these double slit experiments the outcome was arbitrary. A particle traveled at a given speed through a given slit toward a given screen. Where it fell on the screen was affected by a second slit through which it did not pass. As far as the particle was concerned, the identical conditions produced nonidentical results.

The uncertainty principle demonstrated that we cannot measure the present exactly. Quantum physics, and particularly the double slit experiment, demonstrate that even if we could measure all aspects of the present with an error margin of zero, the future would not be predictable. Contrary to all we learned in high school physics, the law of nature known as cause and effect is not a law. It is only a theory. And now, at the quantum level, it is a theory that has been proven to be wrong.

Since identical initial conditions do not produce identical results, the present condition of the universe does not determine the future of the universe. Notwithstanding the ever-present possibility that we may discover the causes underlying phenomena such as these, as we currently understand the world, free will has physics on its side. Does biology also allow us the freedom of choice?

THE BIOLOGY OF FREE WILL

Every cell (with the exception of a very few highly specialized cells) of every body has information within it to reconstruct an entire body. That information is packaged in the double helix of DNA. The efficiency of DNA as a carrier of data is so great that if all the information held within all the libraries of the world (about 10^{18} bits of data) were programmed onto DNA, that information would fit on about 1 percent of the head of a pin.[2] Each cell of our bodies has approximately three billion bits of data coiled within DNA weighing trillionths of a gram.

The DNA of our bodies contains a massive amount of preprogrammed biological information. Though the physics of the universe is not deterministic, what of our biology? Our DNA is a fixed package. We cannot choose the color of our eyes or the color of our skin. Can we choose our emotional temperament or our sexual orientation? Perhaps these characteristics are also controlled by our genes.

The comprehensive role of our genes in our social disposition is a topic still open for debate. However, certain conclusions are absolute. Genes present tendency. They do not dictate our actions.[3–5]

Fraternal twins have in common the same fraction of genes, about half, as do all biological siblings. All develop from a separate egg and separate sperm. When one of a pair of fraternal twins is homosexual, there is a 20 percent chance that both twins are homosexual. Only 10 percent of nontwin pairs of brothers are both homosexual. If genes dictated this trait, the percentage would be the same for both fraternal twins and nontwin brothers since both types of siblings have the same fraction of identical genes.

The occurrence of identical twins is a fortunate quirk of nature. Identical twins form when the egg during its initial divisions following fertilization separates and produces two embryos. As such, identical twins start life with genes from the same egg and same sperm. If genes are dictators of our social tendencies, identical twins should be socially identical. But they are not. In the trait of sexuality, if one of a pair of identical twins is homosexual, there is only a 50 percent chance that the other will also be homosexual.

It is not certain if the 50 percent sexual concordance for identical twins and the 20 percent concordance for fraternal twins are the result of nature (their DNA) or nurture (their social environment). From these data, however, it is certain that while DNA may produce a tendency, DNA does *not* dictate.

The concept of tendency is what biblical morality is all about. Wherever there exists a natural human propensity to an act that is counterproductive either to the individual or the society, there is a biblical command regulating that aspect of life. People have the inclination to cheat in business. Cheating comes in many forms. The Bible forbids them all, describing cheating as an abomination (Deut. 25:13–16). A person may smile at being a workaholic, but the Bible says that one day in seven the drive to transform the material world must be put on hold, and in its place we are to confront the act of simply being, focusing on self and family. If these traits were biologically determined and not alterable, there would be no biblical injunctions to regulate them.

THE THEOLOGY OF FREE WILL

Physics and biology allow us the right to free will. Ironically, it is theology that seems to present the ultimate stumbling block. Among the most ancient biblical writings we are told that God knows the future. Twice in the Talmud[6,7] we read that everything is foreseen by God; nonetheless God grants free will.

If the omnipotent God indeed knows the future, then how, the skeptic asks, can we have free will? The end is already known to God even if we poor mortals do not know it.

The "believer" replies with the reserve of a saint: "Why, how very simple it all is. You see, God is outside of time." That is the kind of statement that pleases the believer but raises the hackles on the neck of a skeptic.

This centuries-old debate was never neatly resolved. Science has finally provided the solution to this theological paradox.

Creation of the universe from absolute and complete nothing marked the beginning of space, time, and matter. Theology has held that position for over three thousand years. Cosmology in the last decade or so has come to agree. These three parameters are characteristics of our universe, not of the Creator. Just as the biblical God is not composed of space or matter, God is also not bound by time. God *is* outside of time. And being outside of time means to exist in an "eternal or unending now," an eternal present that includes past, present, and future simultaneously.

It is my goal to investigate just this reality: that in one reference frame, all times and all events that pass during those times exist *simultaneously* while in another reference frame, those same events are separated by time with a past that has occurred and a future that is yet to come.

There is a bit of graffiti quoted by the renowned physicists John Archibald Wheeler and Edwin Taylor that summarizes this thought nicely: "Time is nature's way to keep everything from happening all at once."[8]

The freedom with which we choose our future is neither absolute nor equal. If we are born handsome, rich, and brilliant we have many more options open to us than if we are born ugly, poor, and stupid. The slings and arrows of fortune, the chance that puts us in a certain environment at a certain time defines the frame from within which Hamlet, and all of us, act out our individual lives.

Yet many prophetic passages in the Bible seem to contradict the concept that the future even to a limited extent is ours to choose:

"And the Eternal said to Abraham: Know for certain that your progeny shall be a stranger in a land that is not theirs and they shall serve
them and they shall afflict them four hundred years . . ." (Gen. 15:13).

The text claims it is "certain" Abraham's progeny will live as exiles.
From Genesis through to Deuteronomy, implications of a foreseen
future are repeated: "And the Eternal said to Moses: Behold you are
to sleep with your fathers but this people shall arise and go astray
after gods of the strangers of the land to which it goes" (Deut. 31:16).

Was four hundred years of abject slavery inevitable for Abraham's
progeny, or was idol worship indelibly written into the Israelite
future as they were about to enter the Promised Land? Apparently
not. Those predicted events occurred only in accord with the actions
of the people. Hence it is written: "And now, if you will hearken to
My voice . . ." (Ex. 19:5). And again, "And it shall be if you fully
hearken to My commandments . . ." (Deut. 11:13). The course of
events is always conditional on the "if," the choice of the people to
follow or abandon the requirements of God. Biblical chronology and
ancient commentary disclose that the Israelites were slaves in Egypt
for just over two hundred years. Where are the missing two hundred
additional years of servitude?

Four hundred years passed from the birth of Isaac, Abraham and
Sarah's child (Gen. 21), to the Israelite exodus from Egypt (Ex. 12).
The land of Canaan, where Abraham and his progeny lived, suffered
severe famines both in the time of Abraham and of Isaac (Gen. 12:10;
Gen. 26:1). This forced them to journey from Canaan in search of
pasture. They stayed but did not settle in these foreign regions, always
returning to Canaan after the famine ceased. When famine came to
the land of Canaan (Gen. 41:54) during Jacob's (the son of Isaac) old
age, the entire clan left Canaan and settled in Egypt: "And Israel settled in the land of Egypt, in the land of Goshen, and got possessions
there and were fruitful and multiplied greatly" (Gen. 47:27).

God had sent Abraham out of Mesopotamia to settle in Canaan
(Gen. 12:5–7). All the time that his progeny remained true to that
charge, they were strangers in the land of Canaan, a land not yet
theirs, but they were free from bondage. Jacob's children changed the
pattern, abandoned Canaan, and *settled* in the land of Egypt." In
doing so they set the stage for their eventual enslavement in that for-

eign land. Had Abraham or Isaac settled in the lands to which they journeyed during famine, much more of the four hundred years might have been spent in bondage.

The example, par excellence, of choice determining the future is found in the biblical book of Esther. The Persian king, Ahashverosh, has chosen Esther to be queen of his vast realm. Esther's uncle, Mordecai, having discovered that the king's first minister has planned the destruction of the Jews, tells Esther to inform the king of this heinous scheme. Esther protests that no one, not even she, is allowed to go to the king without first being called for. Mordecai replies: "Do not think in your heart that you shall escape in the king's house any more than all the other Jews. For if you remain silent at this time, then relief and deliverance shall arise from another place and you and your father's house will be lost" (Esther 4:13, 14).

Esther obeyed Mordecai and so the book of Esther is read today on the holiday of Purim as celebration of the redemption. Had Esther chosen to remain silent, we might be reading the book of Yael or Hadas or Hannah. Esther would be forgotten.

The Bible makes it clear. Our choices affect our futures.

THE MEANING OF GOD BEING "OUTSIDE OF TIME"

We have proven that neither the physics of nature nor the genes of our bodies fix the future. We have seen that the Bible itself confirms that choice shapes the flow of events. If this is so, how does the Creator know our future even before we choose it?

The subtlety in the argument is that we are dealing with two frames of reference, one within and one without the flow of time. For the Creator, being outside of time, a flow of events has no meaning. There is no future in the sense of what will "eventually" happen. The future and the past are in the present. An Eternal Now pervades, like a cloud containing all times, not in a linear progression, but in simultaneity.

The concept of an Eternal Now is implied in the explicit four-letter name of God (Ex. 3:14). In the Hebrew, the spelling includes the letters of the verb "to be" in its three tenses: I was, I am, I will be. The past, present, and future are all contained within the Eternal.[9]

Einstein's discovery of the laws of relativity revealed the astonishing fact that dimensions of space, time, and matter are ever changing and always dependent upon the way in which they are observed. The only constant in our universe is the speed of light (approximately 300,000,000 meters per second in a vacuum).

Einstein theorized and later experiments proved that the faster one travels relative to another object, the slower time flows for the traveler relative to the flow of time measured by the stationary observer. At the speed of light (the highest speed attainable in our universe), time ceases to flow altogether. The time of all events becomes compressed into the present, an unending now. The laws of relativity have changed timeless existence from a theological claim to a physical reality.

For millennia, the stars have been a subject of fascination. Astronomers map the seemingly unchanging positions of the constellations, measure the characteristics of stellar light, and photograph the clusters of galaxies held motionless as if stopped in the midst of a graceful dance. With these data, they formulate theories which describe the history of our universe.

At the Las Campanas Stellar Observatory atop an 8,000-foot mountain in northern Chile, the night of 23 February 1987 started in routine manner. Ian Shelton and his assistant, Oscar Duhalde, were using the observatory's telescope to photograph stars in the Large Magellanic Cloud, a galaxy seen only in the southern hemisphere. By three in the morning, Shelton was about to call it a night. A final photographic plate that had been exposed to the stars for an extended period was to be developed and the night's work would be done. The long exposure allowed the photographic film to accumulate faint light from distant stars not visible to the unaided eye.

As Shelton watched the images appear on the developing negative, an exceptionally large spot, one not present on any of the previous plates, appeared. Was it a flaw in the film? A spot of that magnitude would be visible without the aid of a telescope. He went outside to see for himself. And there it was. A bright star where only a faint speck of light had been. A distant star had exploded producing a spectacular burst of light, a supernova. Its glow had just reached Earth. It was to be designated supernova 1987A.

The star that had exploded was familiar to astronomers. It was located 170,000 light years from Earth. That seems far and it is. But in astronomical terms, it is barely around the corner. It was the closest supernova that had occurred since the development of large telescopes. This burst of light, during the years since 1987, has provided a unique opportunity to study the making of the elements within the residue of that star. We and all our solar system are composed of elements found in star dust such as this.

The light of that exploding star had started its journey through space 170,000 (Earth) years before Ian first saw it. For all those 170,000 years, the secret of the explosion was locked in this burst of photons. Objects closer to the supernova than Earth "learned" of it before we did as the pulse of light passed by them. Had there been intelligent creatures closer than we, they could not have raced ahead of the light and informed us of the supernova, for nothing can travel faster than light.

For 170,000 years, the light of the supernova sped silently through space, bursting out in all directions, a part of it heading toward the place where the Earth would be at three in the morning on 23 February 1987.

At the moment of the explosion, had a Neanderthal hominid gazed at the heavens, he would have seen nothing unusual. The light of the supernova was 170,000 light years distant. After 150,000 more years, Cro-Magnon creatures were making tools and burying their dead, but information of the supernova had yet to reach Earth. Almost six thousand years ago, the *neshama,* the spirit of human life, was instilled in humankind. The image of the Eternal Creator was now present on Earth. Writing was invented and for the first time history was recorded in the form of pictographs. Civilization bloomed. But still there was no knowledge of the event. Through the frigid depths of space the exploding light continued its silent journey.

Some five thousand years ago, the early Bronze Age began. Another fifteen hundred years passed and the alphabet was invented, just one century before the Torah was to be written down at Sinai. For those who searched the skies, there was no sign of the star's explosion. The Israelite exodus from Egypt, the building and destruction of two Israelite temples in Jerusalem, the industrial revolution, the Holocaust, all passed, and the age of space travel and the information revolution dawned. And still the light of the supernova sped silently and secretly through space. And then without warning, on the night of 23 February 1987, it arrived.

On Earth 170,000 years had passed. Tribal villages had become metropolises and the progeny of shepherds had learned to walk in space. Had an imaginary you, not the you here on Earth, but a consciousness devoid of all material aspects, traveled through space in an imaginary massless space ship racing at the speed of light, traveling alongside those photons of the supernova for the 170,000 Earth years that were required for the light to reach us, how much time would this ethereal you have experienced? How many ticks would your clock have made?

The startling, almost incomprehensible answer to this question is: zero. No time would have passed. Not a few years, not a few hours, or a few seconds. Zero time. The difference in the perception of the flow of time at the speed of light is not a *quantitative* difference from a lot of time (170,000 years) to a much shorter time, how ever short that period might be. The difference in the flow of time is a *qualitative* difference, the difference between our existence where all events occur through an unceasing temporally linear flow and an existence in which time does not exist. From that perspective, all the developments that took place during the 170,000 years occurred simultaneously. Past, present, and future had blended into an eternal, ever-present, unending Now. Light, you see, is outside of time, a fact of nature proven in thousands of experiments at hundreds of universities.

I don't pretend to understand how tomorrow and next year can exist simultaneously with today and yesterday. But at the speed of light they actually and rigorously do. Time does not pass.

The biblical claim that the Creator, existing outside of time, knows the ending at its beginning is not because the future has already physically occurred within our realm of time, space, and matter. Einstein showed us, in the flow of light, the corollary of the Eternal Now: I was, I am, I will be.

LIGHT: THE LINK WITH A TIMELESS ETERNITY

It is highly significant that light was the first creation of the universe. Light, existing outside of time and space, is the metaphysical link between the timeless eternity that preceded our universe and the world of time space and matter within which we live.

Light, as with all light-like radiations (the photons of gamma rays, X rays, light, microwaves, etc.), can abandon the ethereal timeless realm of energy and become matter. In doing so, it enters the domain of time and space. Einstein's famous formula, $E = mc^2$, teaches that light and matter are two forms of the same thing—energy. Photons are the ethereal form of energy and matter is the tangible, condensed form. An analogy might be steam and ice being two forms of water.

This link between the eternal and the temporal finds its parallel in the biblical Sabbath. The first holiness in the Bible is neither a place nor an object. It is the intangibility of a time, the Sabbath day, made separate by rest (Gen. 2:3): As Erich Fromm wrote in *The Forgotten Language,* "Rest is a state of peace between man and nature."

Shakespeare concludes Hamlet's famous soliloquy: "To sleep, perchance to dream. Ay, there's the rub. For in that sleep of death what dreams may come when we have shuffled off this mortal coil . . . that dread of something after death, the undiscovered country from whose bourn no traveler returns, puzzles the will and makes us rather bear those ills we have than fly to others that we know not of. Thus conscience does make cowards of us all" (Hamlet 3:1).

As Shakespeare's Hamlet so eloquently insists, the grave may offer no refuge. Although the Creator may know the future, we are responsible for our choices and the actions that result therefrom. Even the most callous of us senses that responsibility.

CHAPTER II

Why Bad
(*and* Good)
Things
Happen

Christopher Marlowe wrote in The Jew of Malta, *"There is* no sin but ignorance." Ignorance is the breeding ground of error. And error is the source of sin. In biblical Hebrew, the generic word for sin is *het*. It means to err, to miss the mark. It does not mean to do evil.

Though *het*, error, is not evil in itself, it may bring evil in its wake.

> *And the Adam knew Eve his wife, and she became pregnant and gave birth to Cain saying, I have acquired a man from the Eternal. And again she gave birth, to his brother, Abel. And Abel was a shepherd and Cain was a worker of the soil. And in time it came to pass that Cain brought from the fruit of the soil an offering to the Eternal. And Abel, he also brought from the first born of the flock and the fat parts thereof and the Eternal accepted Abel and his offering; but to Cain and his offering the Eternal did not accept and Cain was very angry. . . . And it came to pass when they [Cain and Abel] were in the field that Cain rose up against Abel his brother and murdered him. (Gen. 4:1–5, 8)*

In a world according to the Bible, a world in which there is space for free will, bad things are going to happen and they are going to happen even to good people. By the fourth chapter of Genesis this is firmly established. Cain has murdered Abel.

An omnipotent Creator could stop the Cains before they acted, but that would be inconsistent with free will. As long as choice is ours, the possibility for evil, often unintentionally inflicted, exists.

Few people act knowingly with evil design. Most often an erroneous perception of what will result in maximum pleasure provides the motivation behind the error that leads to evil and the tragedy that so frequently accompanies evil. But error need not be a partner to our choices. In the first book of the Bible prior to Cain murdering his brother, we are told explicitly that *het* is not an inherent part of human nature: "And the Eternal said to Cain why are you angry? . . . If you do well you shall be accepted but if you do not well, *het* [error] crouches at your door and to you is its desire but you can rule over it" (Gen. 4: 6–7).

In the next verse the murder occurs. This episode provides a unique insight into the biblical understanding of the human psyche. *Het* is not something with which we are endowed like an unwanted inheritance or a genetic defect. The norm is having goodness in the world. That is the message in having the first humans placed in the Garden of Eden, the biblical version of paradise. Error is the aberration. Unfortunately, as the Bible makes clear in Adam and Eve's expulsion from the Garden, error separates us from the pleasure of "Eden." That separation in biblical language is known as the hiding of God's face. "They will forsake me. . . . Then My anger shall be kindled against them on that day and I will forsake them and *I will hide My face* from them . . . and evil and many troubles shall come upon them" (Deut. 31:16, 17). With God hidden, the world assumes free reign and the potential for troubles increases.

During September 1991, before 250,000 people in Central Park, New York City, Billy Graham endorsed this 3,300-year-old understanding of evil. Occasionally characterized as God's traveling salesman, he acknowledged that he had changed his mind about divine punishment: "I used to think of it as Dante's Hell. Now I think of it more like separation from God."[1]

Cain's punishment for murdering Abel was the withdrawal of God's presence: "And Cain said: My punishment is greater than I can bear. . . . From Thy face I shall be hid" (Gen. 4:13,14).

Criminal trial lawyer Alan Dershowitz remarked to me that God had coddled Cain. For Dershowitz, the proper punishment for Cain's heinous murder of his brother would have been the ultimate, the death penalty. The Bible sees things from a different perspective. In biblical justice, Cain suffered a fate worse than death: the enforced separation from God. He was denied any awareness of the transcendent unity that pervades all existence. He had no hint of a larger purpose other than day-to-day survival, a living death.

NATURE'S ROLE IN TRAGEDY

If all suffering were attributable to human agency, then free will and our inherent ability to err might make its presence more readily logical. Unfortunately, the blind forces of nature lie behind much human grief.

An earthquake shakes a bridge from its foundation, dropping it onto a crowded bus passing beneath. A chance cosmic ray smashes into an ovum, produces a free radical which in its natural drive to establish electrical balance tears and mutates a chromosome. As a result, a crippled child is born. The same Creator that produces the beauty of a sunrise and the colors of a flower must be credited with these horrors as well.

Instinctively we might recoil at such a Creator. But instincts are not always the best guides in complex situations.

We might speculate that each person on the bus was there by design. The data-handling power of a supercomputer could produce the perturbations required to arrange the lives of selected persons so that they alone would be on that ill-fated bus. Obviously an omnipotent Creator could do the same.

Such divine planning is possible. I am not certain that biblical religion demands it to be continuously implemented. The famines that plagued the biblical patriarchs Abraham, Isaac, and Jacob throughout the Book of Genesis might have been God's deliberate work. But the text does not say so. They also may have been one of the many forms by which nature, with its leeway, constantly challenges us. It is for us to learn how to react to the bad as well as to the good even if we cannot understand the purpose of either (Deut. 28:47; Deut. 31:20; Hosea 13:6). In a natural universe, earthquakes and cosmic rays as well as rainbows and blue skies are intrinsic parts of the scheme.

The slow churning of Earth's iron-rich, molten outer core sends energy toward its surface. As a result, continents glide slowly across the globe. Occasionally as a continent slips along, moving a centimeter or so each year, it snags on an adjacent land mass. The pressure accumulates until in the jolt of an earthquake, the snag breaks and the continent continues its journey.

Like the honey and the sting of the honeybee, we need the Earth's molten core, though we would prefer to avoid the quakes it produces.

For life, a planet must be close enough to a star (the Sun) to receive the heat needed to keep water liquid. At that modest distance, massive doses of cosmic radiation accompany the warming rays of the star. The same motion of the molten iron core that propels continental drift and in the process produces occasional earthquakes continuously shields us from that radiation. The moving slurry of iron and rock deep within Earth produces and maintains a magnetic field that surrounds it. The force of that field deflects much of the cosmic radiation that would otherwise bathe the planet's surface with lethal doses of energy. Stopping the motion of the Earth's core would put an end to earthquakes, but it would also eliminate the protective magnetic field.

The biblical Creator has the ability to form stars without lethal radiations. But they would not be natural. They would offer absolute testimony to the existence of the Creator.

Mutations of the genome only rarely produce the tragedy of a crippled child. Most often these errors are corrected or discarded long before birth.[2] The rareness of deformity attests to the efficiency of our protective genetic program. But occasionally a random mutation slips through.

Obviously, an omnipotent Creator could remove all randomness from nature. Crippled children and hereditary diseases would be no more. But the price would be too high. Without some degree of randomness, all events and all choices in the universe would be totally predetermined by unyielding laws of nature, the physics and chemistry of all reactions. We would be mere robots. Our every thought and action would be fixed by the immediately preceding chemistry of our bodies and the conditions of our environment. The future would be totally controlled by the past.

Freedom in nature—so that not every stellar system is a life-nurturing solar system—and freedom of will—so that a given stimulus produces a variety of responses—are traits of the divine contraction, the *tsimtsum,* which brought our universe into existence. Whether *tsimtsum* is divinely essential in universe formation or a deliberately chosen aspect of the design of our universe is a question we cannot answer. It is, however, the reality of our existence.

In nature, free will and the potential for tragedy go hand-in-hand.

The world around us presents a web of facts from which we learn. Seeing through this web and discovering justice and righteousness (Deut. 6:18) is the challenge. Considering that the human brain has the capacity to store the information contained in a fifty-million-volume encyclopedia, we ought to be sufficiently wise to succeed at the task.

Unfortunately, on an absolute level, we do not observe the world as it is. Before we can assimilate a "fact," the external objective reality of our world must pass through a series of biological and mental filters. The data are no longer data. They are personalized summaries of the external facts. As the saying goes, beauty is in the eye of the beholder. So is color, along with a host of other sensory perceptions.[3] No wonder the spies that Moses sent to search out Canaan prior to the Israelite entry into that land came back with a report that they were like grasshoppers relative to the "giants" that populated Canaan (Numbers 13:32,33). Objective reality evades our human grasp.

The human brain is layered, with the outermost part, the cerebral cortex, accounting for over 60 percent of the total brain mass. The cortex gives us the ability for critical analysis. Here is where most of those fifty million volumes of our mental encyclopedia are stored. The cerebral cortex's capability for logic might come up with truth every time were it not that beneath it lies the limbic layer of the brain. This layer feeds the cortex with a range of emotions, feelings, prejudices, and lusts that color the objectivity of our "analysis."

Reason might be able to filter out the emotions and come up with absolute truth if the limbic were the end of the tale. But it is not. Just below the limbic lies the r-complex. The aggression, territoriality, and greed that influence our thoughts arise here.

Our inherent subjectivity makes Spinoza's concept of absolute justice derived by reason an illusion. Lawyer Dershowitz remarked that his clients consistently managed to justify their acts to their own satisfaction, even if not to the courts'.

If we could construct a being without these lower levels of the brain, reason alone might produce a just and righteous world. Truth without the mask of self-interest could shine through. For millennia, scholars have pondered the characteristics of such a pure being and have concluded that not even this would be the solution. Without the lusts for accomplishment and control which originate in the more primitive parts of the brain, we might never have the drive it takes to build a world, to have "dominion over the earth" (Gen. 1:26). As with the forbidden fruit, good and evil come together.

In his popular book, *When Bad Things Happen to Good People,*[4] Harold Kushner claimed that God is limited. Kushner suggested that there are regions into which God's power cannot extend and that is why bad happens. Biblically there is no foundation for such an idea. Famines and holocausts do not occur because God is limited. The Bible itself informs us repeatedly that through *tsimtsum* and the occasional hiding of the divine presence, God allows them to happen as part of the divine scheme. What seems to be divine indifference lies not in some inherent limit to the Creator. Rather it is the foundation of our free will.

Animals have choice because, according to the Bible, animals have a *nefesh,* and the *nefesh* provides the freedom to process and evaluate information. Biblically an animal is defined as *nefesh hiyah*—a living *nefesh* (Gen. 1:20). In place of the generations required for random mutations to produce meaningful changes in the information held on DNA, within milliseconds the *nefesh* can deliberately alter mentally stored information. The human brain has ten thousand times the capacity for information as has the human genome. The brain has liberated animals from the tyranny of their DNA.

The *nefesh,* by analogy, might be described as a computer having a self-correcting program set to maximize pleasure and survival. (I use the computer as a metaphor only, with no judgmental intent of a direct similarity between aspects of life and computers.) Animals investigate the environment and from their experiences rapidly learn to optimize their pleasure and their chances of survival. With cunning, a host of animals learn to pass through a maze to gain a reward on the far side. Anyone who has raised a pet knows that many animals communicate via sounds that can be described as primitive, noncomplex language. They have sufficient awareness of self to recognize and herd with their own species in the wild. City-bred dogs bark at and play-wrestle with other dogs. Even though there may be close to an order of magnitude difference in size, a Great Dane realizes that the Jack Russell terrier is a tiny version of himself. I have yet to see a dog come upon a stray cat and play-wrestle with it. Cats, they realize, are not dogs.

Animals learn and animals choose. But their options and inclinations are limited according to the inputs the *nefesh* gets from the genes and from the body. Avoid pain, seek food, produce, protect, and nurture the young, seek pleasure.

Humans also are strongly driven by these basic desires of survival and pleasure. We and all animals are pleasure seekers. But humans have a source of pleasure not evident in other animals. It arises from the *neshama,* our link to an all-encompassing unity that underlies what superficially appears to be a diverse and multifaceted universe. The *neshama* whispers to us of a pleasure that transcends our limited physical existence.

The decision-making program of the human *nefesh* now has two sources of information to consider as it strives for pleasure: the desires and needs of the body and the spiritual goals of the *neshama.* How I choose to achieve my pleasure determines the quality of my person.

When I exercise my will, I choose from within an ever changing window of possibilities built from my cumulating knowledge and experiences. No two people have the identical window of choice because no two people have the identical physical and social histories. As we increase our knowledge and expand our experiences, the boundaries of our windows shift. Quantum mechanics has proven that the visible world is not a direct extension of the subatomic world from which it is constructed. By this it has laid basis for the biblical concept that the mind is not a mere extension of neurochemical reactions of the brain. QM does something much more important than destroy determinism—it provides the opening *and the mechanism* for choice.

Through speech and body activity, the brain translates into physical manifestation the mind's thoughts. The conscious mind draws its inputs both from biochemically determined neurological activity and from what appears to be a partly random, nondeterministic surfacing of subconscious thoughts into the conscious. Recall that the eye responds to a single quantum of light (the photon) as does almost all matter. The energy transfers resulting from individual photons impacting on orbital electrons are what yield the secondary radiations we see as color. For the eye, the photon's impact on the retina starts the neurological signal that carries information to the interior brain (the eye being an exterior extension of the brain). With this in mind (no pun intended, though it is worthy of reflection), a quantum mechanical (and therefore nondeterministic) concept of thought finds foundation.

What might be described as a quantum wave function includes the range of an individual's subconscious information (this being the subconscious portion of the window of choice). The wave, in a manner that is not precisely predictable, then collapses, focusing on one concept and that surfaces as conscious thought. But the collapse, while not deterministic, is also not totally random. It is highly skewed by the individual's history. If tension and anxiety underlie the experiences fed to the mind, then tension and anxiety will surface in thought. The rhythm of the biblical sabbath opens guaranteed shelf space for repose in our mental libraries. What Louis Pasteur said concerning scientific observation is true for all of life: chance favors only the mind that is prepared.

The only unalterable destiny in life is that life itself is a continual crossroads of choice and the driving force behind that choice is the search for what we perceive as pleasure. Mistaking short-term gratification for transcendental pleasure is something like confusing infatuation with love. How much misery has been founded on that error. But then, immediate pleasure is so tempting.

> *And the Lord God planted a garden eastward in Eden, and there he put Adam. . . . And the Lord God made grow every tree that is pleasant to the sight and good for food; the tree of life within the garden and also the tree of the knowledge of good and evil. . . . And the Lord God commanded Adam saying of every tree of the garden you may certainly eat but of the tree of the knowledge of good and evil you may not eat, for in the day that you eat of it you shall surely die. . . . When the woman [Eve] saw that the tree was good for food and that it was tempting to the eyes and suited to make one wise, she took from its fruit and ate and gave to her man with her and he ate: (Genesis 2:8,9,16–17; 3:6)*

There were all kinds of wonderful trees in the Garden, even one that gave eternal life and another that gave knowledge of good and evil. Now their Master and Maker and Creator had permitted Adam and Eve to eat from *every* tree therein. Well, almost every tree. There was one that was forbidden: the tree of knowledge of good and evil.

And the punishment for eating from that tree was death. Luckily there was a loophole. To get around the promised calamity of death if they transgressed the prohibition, they could eat first from the tree of life and then later from the tree of knowledge. The choice was between eating from a tree that would grant life or from a tree that would bring death. In basic terms the choice was between life and death, not between good and evil.

Just prior to the completion of the Torah, the Bible defines the choice that our free will faces. In the entire five books of Moses, this is the *only* place where we are instructed that we must make a choice.

> *I call heaven and earth to witness against you this day, that I have set before you life and death, the blessing and the curse; therefore choose life that you may live, you and your progeny. (Deut. 30:19)*

That admonishment in Deuteronomy occurred 2,400 years after Adam and Eve's expulsion from Eden, and 3,000 years before Shakespeare's Hamlet first uttered his famous soliloquy. But humanity had not, and has not, changed throughout the generations. The choice with which we battle is between life and death, not between doing what is good or what is evil. Most of us think that what we are doing is for the good, personally or for our families or for our nation, even if others see our actions otherwise. It's just that sometimes we err and end up in a mess. As Shakespeare wrote so wisely, "Things bad begun make strong themselves by ill" (Macbeth 3:25).

With the stakes of our choices so high, life and death, it would seem prudent to make every effort to choose correctly. That takes knowledge, which fortunately is available. We have only to seek it.

CHAPTER 12

Bread from
Earth:
A Universe
Tuned
for Life

A land where you will eat bread without scarceness. . . . You shall eat and be satisfied and bless the Eternal your God for the good land which he has given you. (Deut. 8:9,10)

Western religions have an age-old tradition of blessing food. Perhaps all religions have this concept of offering thanks for divine beneficence as the ultimate source of sustenance. While doing research in the hinterlands of Thailand, I trailed behind white-robed monks with shaven heads as they walked from hut to hut early each morning with empty wooden bowls in hand, asking for a gift of rice. In one act they taught the donor to offer charity and to be grateful for the bounty they all enjoyed.

"For what we are about to receive, Lord make us truly grateful." This, with slight variations, is the typical Christian pre-meal blessing. Judaism has a formalized statement: "Blessed art thou Lord our God, King of the universe, who bringest forth bread from the earth."

Is it the King of the universe who brings forth bread from the earth? Or, as a skeptic might ask, is it the farmer who tilled the soil and the baker who formed the loaf? In this chapter we examine the so-called anthropic argument—the idea that the physical, chemical, and biological laws of nature are so fine-tuned that they could not have occurred by chance. By this line of reasoning, if indeed we are not just one of an infinite number of universes with an infinite number of differing natures and types of life, then only an intentionally created universe could have produced something as unlikely as bread.

What does it take to make a loaf of bread? When I ask this question in my classes, the list usually begins with flour, water, yeast, sugar. With some professorial prompting soil, sunlight (an extraterrestrial input), a stove (engineering), and recipe (intellect) are added. We're getting more complex. Still more prompting elicits a desire for the bread (choice), a nurturing climate, and finally, in the cosmic sense, a nurturing universe. To make a loaf of bread you need a very special universe. Our universe has been special since its inception.

In the beginning, from total and absolute nothing, the Creator brought forth a substance so thin it had no corporeality, but that substanceless substance could take on form. This was the only physical creation. Now this creation was a very small point and from this all things that ever were or will be formed. . . . If you will merit and understand the secret of the first word, Beresheet *[In the beginning], you will know why the Jerusalem translation [of Gen. 1:1] is "With wisdom God created the heavens and the earth." But our knowledge of it is less than a drop in the vast ocean.*

Thus wrote the kabalist, Nahmanides, in his commentary on the opening of Genesis. More fundamental than matter, according to the kabalistic interpretation of Genesis, is a substance so thin it lacks corporeality. Yet this substance could turn into matter as we know it. Is there a scientific suggestion for a thin, noncorporeal substance that is the source of all matter?

Confirmation of this illogical reality was Einstein's mind-boggling proof that matter is energy in its condensed form. Intuitively, this is ridiculous. And so its initial label as theory, not law, lasted for decades. What does energy have to do with matter? How is the photon of a light ray related to a pencil or a rock? The idea that a light ray could actually change into something material was beyond the pale of reason. So we thought. We were wrong.

And yet more fundamental than energy, according to kabalah, is wisdom: "With wisdom God created the heavens and the earth." "By the word of the Eternal the heavens were made" (Psalm 33:6), and a thousand years later, "In the beginning was the word" (John 1:1). According to the Bible, the entire physical structure of our vast universe is a manifestation, a concretization, of wisdom. On this concept, science is (as of now) silent. Considering the information explosion that we are experiencing, a few more decades may be all that is required for science to confirm the next step in the biblical theory of our origins, that wisdom is the unifying base of energy.

Pantheism comes close to grasping this insight, but stops short of the crucial final step. It sees the universe as an integrated whole, a unity operating through the diverse laws of nature. This is a brilliant insight. But then it claims that these laws and forces active in the universe are all that there is. If there is a god, it is these laws. The difference between pantheism and monotheism is that biblically, the laws of nature are understood as a projected manifestation of an infinite wisdom that transcends the physical universe, within which the physical universe dwells, and of which the physical universe is composed.

Kabalah claims that the wisdom of the universe is made manifest in a loaf of bread.

In the second book of the Bible, Exodus 25, there is a detailed description of the Tabernacle that accompanied the Israelites during their forty-year sojourn in the desert. Its vessels are always listed in the same order. First comes the ark holding the Ten Commandments, next the table upon which each sabbath twelve fresh loaves of bread were placed. Then comes the seven-branched candelabrum and, still later, the sacrificial altar.

The table is always listed second, immediately following the ark. One might expect that the most important item following revelation (as represented by the ark) would be the spirituality of the candelabrum's flame or the submission of sacrifice on the altar. Biblical religion sees priorities otherwise: first the centrality of the Creator's word, then a table with twelve loaves of bread. Bread is the source "that sustains the human's heart" (Ps. 104:15). The table upon which these loaves were placed is the symbol of nature's bounty. The Word would mean little if we lacked the sustenance to enjoy it.

The details of the ark's construction require twenty-three verses; that of the table, eight verses. Among their many components, both have a gold rim or crown *(zer zehav)* defining their borders. Of the hundreds of words these descriptive verses comprise, kabalah selects two, "gold rim," and the gold rim of the table, not of the ark, to tell us how the Bible views nature: "In this explanation lies the secret of the Table [in the tabernacle]. For since the time that the world came into existence, God's blessing did not create something from nothing; however, the world follows its natural course."[1,2]

God's role in nature is made manifest in the laws of nature. The universe is filled with material munificence. And nature has produced that abundance. We don't need a table and a few loaves of bread for that information. It's obvious from a glance out the window. However, the reason for that abundance is less obvious.

The table's gold rim informs us of boundary, not of bounty. As the rim marks the border of the symbol of nature's bounty, the table, so too is nature bounded, confined, channeled by a set of laws that yielded from the chaos of the big bang, the consonance of a loaf of bread, and the consciousness to enjoy it. To make a loaf of bread takes a very special universe.

The big bang produced energy, all of it squeezed into a volume possibly smaller than your thumb. It was the only physical creation and it marked the beginning of time, space, and matter. Physics and the Bible are completely in agreement on this. What lies beyond the universe is pure speculation. It may be that nothing, in the most complete meaning of that word, is beyond. There is no vacuum within which the universe hangs. A vacuum implies space and space is part of the creation. The edge of the universe would be an edge with an inside and no outside, a phenomenon that can not be pictured in our minds or by our computers.

From the energy of the creation, all matter that ever was or will be formed. With the wisdom of hindsight, most especially the knowledge of the fact that we along with 10,000 billion billion stars strung gracefully through 100 billion galaxies are here, we can speculate upon how that first matter formed. Looking forward in time from the creation, from the vantage of the big bang itself, the picture would be very different.

Most laypersons are unaware that according to the classical laws of relativity and cosmology, there should be no solid matter in the entire universe. We and the Earth on which we rest and every star and galaxy in the universe are the result of an unlikely quirk of nature.

Einstein's famous formula, $E = mc^2$, tells us that as long as radiant energy (E) is more powerful than a specific threshold needed to make a particle of matter, that energy can change spontaneously and become a particle of nuclear matter (m). This is the source of all matter of the universe.

Yet an absolute and undisputed physical constraint in the formation of particles from radiant energy, the E changing into $m,$ is that the particles form in pairs, a particle and its antiparticle. Matter and antimatter, most fundamentally the subatomic particles known as quarks and antiquarks, are produced in equal amounts. Quarks combine to form protons and neutrons which in turn combine to form the nuclei of all atoms. From them all galaxies, stars, and life are fashioned. Antiquarks combine to form antiprotons and antineutrons.

Unlike protons and neutrons which can exist at all naturally occurring temperatures found in the universe today, from the frigid reaches of space to the infernos at the centers of the most massive stars, quarks are found only at temperatures greater than about 10^{13} degrees Kelvin (ten trillion degrees above absolute zero). Below that temperature quarks are confined in protons and neutrons, and once confined there they stay. Hence the term quark confinement. Such extreme temperatures exist today only in the artificial environs of the most energetic particle accelerators at advanced physics laboratories. Free-existing quarks are not found in nature.

As the universe expanded following the big bang, its temperature decreased in direct proportion to its increase in scale. That is the nature of all expansions. The expansion of freon in a tube in your refrigerator is what cools it. The expansion of air as it rises explains why there is snow at the peak of Mt. Kilimanjaro even though the temperature at the base is a sweltering 100 degrees in the shade. The cooling effect of an expansion is logical. It is the dilution of a given amount of heat in an ever larger volume.

When the universe was ten million million times smaller than it is now, its temperature was above the temperature of quark confinement. Quarks and antiquarks were being formed from energy. It was an unstable world. The nature of a particle and its antiparticle is that they are identical in most aspects, but when they touch they literally explode and self-annihilate in a burst of energy. That is the *m* (of $E = mc^2$) changing back into *E*. In the first fraction of a second after the big bang, quarks and antiquarks were forming and annihilating continuously, returning to energy and then reforming. Energy levels were so high that following an annihilation, new quark pairs were immediately formed. But the universe was expanding and with the expansion came cooling. As the temperature dropped below the threshold level, the remaining pairs of quarks and antiquarks immediately self-destructed. The energy now was below the level at which they could reform. Since matter and antimatter always form in identical amounts, all quarks should have rapidly annihilated and produced a universe filled only with radiant energy and no particles.

And that is what almost happened.

"The obvious question is," *Scientific American* asked in 1993, "How is it that so much matter managed to survive? . . . Why is there something rather than nothing?"[3] No matter should have survived. In theory, the universe today should be matterless, containing only the faint glimmer of radiation so weak it would not even match the energy of a single microwave in an oven.[4–6]

For an exotic, still uncertain reason, infinitesimally more matter than antimatter was produced. We and all the material universe are testimony to that primordial inequality. The difference in pairing was small, one part in ten billion. That is, for each 10,000,000,000 antiparticles, 10,000,000,001 particles formed. As the particles and antiparticles annihilated, that one extra particle in ten billion remained. From those rare "extras," every galaxy, star, and human is composed. Steven Weinberg referred to it as an "embarrassing vagueness . . . the unwelcome necessity of fixing initial conditions, especially the [10] thousand million to one ratio of photons to nuclear particles."[7] "The one part in ten billion excess of matter over antimatter is one of the key initial conditions that determined the future development of the universe."[8]

The source of the surplus is speculative. There might have been a fortuitous "error" in accounting as the pairs were being produced. This is actually allowed in rigorous science. Over very very short time spans, random events such as particle production can occur in disequilibrium, which then almost immediately is reset into equilibrium, particle and antiparticle. There are speculations about ways that this disequilibrium might have been "frozen" into the composition of the universe during the quantum of time prior to establishment of equilibrium.[9]

Alternatively a source of particle production other than direct pair production may have existed. In theory, this could have been the nonsymmetrical decay of a hypothetical supermassive radioactive particle (call it X) and its antiparticle (anti-X) that were present only in the very young, very hot universe. Although X and anti-X were present in identical numbers and, being particle/antiparticle pairs, both must have decayed at precisely the same rate, they might not have decayed via the same decay channels and so they may not have yielded the same end products of the decay. As such, they might have produced slightly more protons and neutrons than antiprotons and antineutrons.[10] Unfortunately this theory cannot be tested. The conditions are no longer available for scientific scrutiny.

There is no reason, a priori, that the symmetry that produces matter and antimatter in identical quantities from energy should have been broken early in our cosmic history. Yet it was. Our presence testifies to that break, but not to how it happened. By the time the universe had expanded to about the size of the solar system, the dominance of matter over antimatter was fixed. It took place in the first one hundred thousandth of a second following the big bang. We, or at least the building blocks from which we would be formed fifteen billion years later, were written into the universe at its earliest moments.

Having broken the symmetry and started the process of forming the elements from that energy and those "few" bits of excess matter, the physics of nuclear stability favored a material universe composed totally of iron. That sounds good for the iron-rich hemoglobin of our blood cells, but of course to make those cells and all the rest of life we need things like carbon and hydrogen and nitrogen and oxygen, to name just a few of the essential ninety-two elements of our universe. Our bodies represent a chemist's shopping list.

Here too, nature filled that list. Tucked within the boundary of the table's gold rim is a quirk of nature, an instability in a transition state between hydrogen (the lightest of elements) and helium (number two in the elemental list). That instability stopped the initial nucleosynthesis at a universe composed approximately of 78 percent hydrogen and 22 percent helium, by weight, or 93 percent hydrogen and 7 percent helium atomic abundances.[11,12]

All this occurred during the first three minutes following the big bang. But none of it would have, had it not been for the extraordinary balance of power played out among the strong nuclear force, the weak nuclear force, and the electromagnetic force. These meshed in just correct proportions so that protons and neutrons drew together and bonded, thus forming the nuclei of atoms. The temperature of the universe was then approximately 300 million degrees Kelvin and its scale 100 million times smaller than at present. Had the balance of nuclear forces been otherwise, the end product of the big bang might have been a jumble of subatomic particles bathed in the radiation of ever weaker photons.

A few hundred thousand years passed. The heat of the big bang became ever more dilute in the continually expanding volume of the universe. The temperature of space fell to below 3,000°K. Atomic nuclei could now capture electrons. At higher temperatures and earlier times, the photon energies were so powerful that they continuously smashed electrons free from their atomic orbits. The universe was a thousand times smaller, and a thousand times hotter, than it is today.

Gravity becomes the dominant force in the universe at distances greater than the size of molecules. This amazing force of nature pulled the primordial hydrogen and helium gases into galaxy-sized clouds. Pockets of gases formed local concentrations. Mutual gravitational attraction among the gas particles then squeezed the individual atoms so tightly together that their nuclei fused. But the fused mass weighed less than the original, separated particles. The lost mass appeared as energy, and that released energy self-ignited a chain reaction, a sort of mammoth H-bomb or a natural controlled-fusion nuclear reactor. It was the birth of a star.

Millions and billions of years passed as the alchemy of nuclear processes under the pressure of a star's core fused lighter elements, each having only one or a few protons and neutrons per nucleus, into the heavier atomic nuclei so essential for life. When a star's entire supply of hydrogen had fused into heavier elements, the star's primary energy source vanished. Like a hot air balloon suddenly quenched with a cold blast, the outer layers of the star imploded and then in rebound exploded, literally spewing the stellar debris into space. The elements of life were now present in the universe as star dust, formed step by step from the energy of the big bang.

Some five billion years ago, approximately ten billion years after the big bang, in this corner of the universe a star formed from a spinning mass of primordial hydrogen mixed with a (literally) healthy dose of recycled star dust. Not all the dust and debris were drawn into the stellar center. As the spinning gas and dust flattened into a disk, aggregates circling the star formed into nine planets and a jumble of asteroids concentrated between the fourth and fifth planets. Tidal forces from the gravity of the immense fifth planet out from the star kept the asteroids from agglomerating into a tenth planet. On the third planet out from the star, sentient beings eventually named the star Sun.

Today the universe is 10 to 18 billion (thousand million) light years in scale. The temperature is a bit less than 3°K. That's approximately -270°C or -450°F. And it's getting colder all the time. In another fifteen or so billion years, the temperature will have halved to 1.5°K.

We live on a planet within a galaxy of a universe made for life. As hostile as the frigid reaches of space may be, it is the vast universe that provided the time and energy to allow stars to develop and then to die in supernovae explosions, thus producing the elements needed for life and then seeding them into space. Those same depths of space shield us from the lethal radiations coming from the supernovae.

Closer to home, the life support system we call the solar system has just the correct distribution of large and medium sized planets to have swept clean most of the space through which Earth must travel. Our star, the Sun, is just the right size to consume its supply of hydrogen and produce energy at a rate that provided the time and conditions for life to form. Our orbit through space, at 150 million kilometers from the Sun, departs from a true circle by only 3 percent. Were it as elliptical as the orbit of Mars, the next planet out, we would alternate between baking when closer to the Sun and freezing when distant. Earth contains just enough internal radioactivity to maintain its iron core in a molten state. This produces the magnetic umbrella that deflects an otherwise lethal dose of solar wind. The volcanic activity driven by this internal heating is just adequate to have released previously stored subterranean waters into our biosphere, making them available for life processes, but not so much volcanism as to shroud our planet in dust. Earth's gravity is strong enough to hold the needed gases of our atmosphere but weak enough to allow lighter noxious gases to escape into space. All this is balanced at just the correct distance from our star so that our biosphere is warm enough to maintain water in its liquid, life-supporting, state, but not so warm that it evaporates away into space.[13] We are not dependent on Earth alone. We are truly children of the cosmos.

A just-right Earth with just the needed gravity, radioactivity, magnetic field, and volcanic activity to support life is located at just the correct distance from the Sun to nurture the inception and development of life.

But Earth should not be where it is. Among the planets circling the Sun, Earth is the oddball. The distribution of matter initially spiraling in toward a central attractor may reach an equilibrium that clusters along what is known as an exponential curve. In this curve, each successive swirl is a given factor farther out than its predecessor. The distances of planets from the Sun fall on an exponential distribution. Each planet is approximately two times farther from the Sun than the preceding planet, except for Earth. Earth should not be where it is.

In millions of kilometers, the distances from the Sun are: Mercury 58, Venus 110, Earth 150, Mars 230, asteroids 440, Jupiter 780, Saturn 1430, Uranus 2880. Neptune and Pluto change position of order due to Pluto's eccentric orbit. Pluto moved closer to the Sun in 1979 and will remain so until 1999. Some force in the past disturbed the orbits of these outer two planets. If we therefore only consider the inner seven planets and the asteroid belt, Earth, at 150 million kilometers from the Sun, does not fit on the exponential distribution. Yet here we are in all our life-giving splendor and awe. A miracle? Perhaps, or just a fortunate quirk of nature.

The renowned Oxford mathematician Roger Penrose quantified the precision needed in nature's quirks for the conditions and energy distribution at the moment of the big bang to have eventually produced an environment suitable for life. The likelihood, or better the unlikelihood, that those initial conditions might produce such a universe is less than one chance out of ten to the power of ten to the power of 123. That is one out of a billion billion billion, etc., repeated more than a billion billion times.[14] Just to speak aloud those billions would require more time than the universe has existed, more time than has passed since the beginning of time!

If the nature of our universe were not so finely balanced, either it would have expanded so fast that matter would not have had the time to agglomerate into galaxies and stars and planets, in which case the universe would now be a mist of dust and ever weaker radiation, or it would have collapsed in less time than required for galaxies to form. Either way, the universe would today, and forever, be lifeless.

Consider gravity. "We do not understand what mechanism generates mass in the basic building blocks of matter." The extraordinary significance of this statement by the president of the Massachusetts Institute of Technology, Charles M. Vest, in his 1995 annual report is not at once evident to the scientifically uninitiated. We do not know why there is gravity! Why does something, anything, and everything have weight? Why do we stay seated in a chair on the floor rather than floating off through space?

Light, traveling and shining and beaming its way through space, exists as a wave and as a particle simultaneously. Wave/particle duality in itself is baffling, but at least amenable to discussion. The electromagnetic field made manifest in a photon of light carries information and force, transferring it from source to receiver. The Sun shines forth photons and eight minutes later, from our perspective, those photons reach us bringing with them information of the color and warmth of the Sun. The Sun has sent us something and we received it.

But gravity? What allows gravity to accomplish its task of holding you in your seat and Earth in orbit around the Sun? We theorize that as light has quanta of energy called photons, so gravity has a gravitational field whose quanta, or energy pockets are called gravitons. But contemplate the difference between light and gravity. Why should Earth sending its gravitons to the Moon and the Moon sending its gravitons to Earth lock the one in orbit about the other? Why should a transfer cause attraction between bulks of matter? Yet there the Moon floats, held by an invisible tether, a force generated by unknown processes with imagined particles. By what unlikely and fine-tuned principle does it work?

Similarly, give thought to the amazing consistency of radioactivity. A billion atoms of the radioactive noble gas radon all have the half period of 3.83 days. That means 3.83 days from now only half a billion atoms will remain and in twice that period only one quarter will remain. But which atoms will radioactively decay now and which later? Some will decay after a second and some only after a century. Yet all have the same 3.83 day half period. Do they decide among themselves? Are the individual atoms aware of how many have already decayed and so how many must decay today? Or more generally stated, is matter in some sense aware of its environment? It often seems so for radioactivity and also for a variety of other occurrences regularly observed in physics laboratories.

And then there is life, human life for example. Its chemistry is just as extraordinarily well tuned as is the physics of the cosmos. Our world on both sides of the divide that separates life from lifelessness is filled with wonder. Each human cell has a double helix library of three billion base pairs providing fifty thousand genes. These three billion base pairs and fifty thousand genes somehow engineer 100 trillion neural connections in the brain—enough points of information to store all the data and information contained in a fifty-million-volume encyclopedia. And then after that, these fifty thousand genes set forth a million fibers in the optic nerves, retinae having ten million pixels per centimeter, some ten billion in all, ten thousand taste buds, ten million nerve endings for smell, cells that exude a chemical come-on to lure an embryo's lengthening neurons from spinal cord to target cell, each one of the millions of target cells attracting the proper nerve for the particular needed function. And all this three-dimensional structure arises somehow from the linear, one-dimensional information contained along the DNA helix.

Only some 3 percent of the human genome appears to actively code for genes. Even if the entire genome were active, there would not be enough points of genetic information (DNA) to make a human. There is not enough information in any mammalian species' DNA to make a member of that species. Only if the information is compressible—that is, only if the minimum amount of directing information required to make a specimen of the morphological class known as mammal is much less than the sum of all the "information" present in the body (the sum of all the proteins, organs, skeleton, etc.)—could, for example, a sentient intelligent human arise from a cell and a string of nucleotides containing three billion bases.

This compression of information would be represented by an algorithm embedded in its DNA. But whence cometh the algorithm? Does not the idea of a parsimonious algorithm conflict with the known reality that the genetic codes of all animals contain large amounts of redundant information and therefore are not manifestly parsimonious? Science, if rooted in the physical, will have the answer for the origin of the compression: chance, a lucky roll of the dice. That was the conclusion of a research article published in *Nature*.[15]

To account for the lack of specific genetic information, the genome may only provide general instructions. But these generalities must be translated into specifics, and that takes information. All this directional information is almost error free, and usually produces viable healthy beings. From where does it arise? We are the embodiment of a synchronous mystery.

A nerve axon extends toward its target. What complexity is hidden in those few words. They assume the orchestration of millions of atoms into a unified action. Within each cell hundreds of thousands of molecules, mostly proteins each containing multiples of thousands of atoms, must be shunted from the site of their production to their point of use. Carrier molecules have learned to recognize the proper cargo and its destination. As a way of supporting arguments for evolution, computer programs blithely show the transition of outer body forms from amoeba to fish to amphibian to reptile to mammal and human. These electronic displays deliberately ignore the intricacy of the molecular functions of each cell and serve only to repress the impossibility of randomness as the driving force behind these life processes. They are in fact an exercise in deception, an insult to adult intelligence.

Over half a century ago, discoveries in quantum physics forced us to relinquish the hopeful rationale that actions in the subatomic world occur in a manner able to be derived by logic alone. Molecular biology, the chemistry of life, has revealed the underlying complexity of the simplest organic processes.[16] It has come to do for the sciences of life what quantum mechanics has already accomplished for physics. The unlikelihood of carbon, liquid water, and a perfect Earth pale by comparison with the unlikelihood of DNA, cellular machinery, and biology's organizational functions.

The Hebrew word for the world is *o-lam*. It also means "hidden" and "eternal." What we see in the universe, even with the best of instruments, is but a hint of the total world. Most of existence is hidden—perhaps, as kabalah states, folded into twenty-six basic and ten general dimensions, only four of which are available for our examination. (String theory makes a similar estimate.) Here is how kabalah makes that calculation:

The Eternal told Moses "this is My name *le o-lam*" [forever hidden] (Ex. 3:15). The letters of the Hebrew alphabet have numerical value. Originally letters were the only symbols used in Hebrew for recording quantity. Aleph, the first letter of the alphabet, was always one; beit, the second letter, two; gimmel, the third, three, and so on. The explicit name of God has four letters, the numerical value of which is twenty-six. The exact pronunciation of this name according to kabalistic tradition is forbidden but sometimes it is spoken of as Je/hov/a. Of the twenty-six dimensions, only four are known. The other twenty-two are closed to our vision, hidden forever. Perhaps within them rests the means by which the Moon is held in orbit around the Earth and a single fertilized cell directs the production of the ten trillion (10^{13}) cells that make up a completed human.

The impact a hidden dimension might have on physical reality is not simple to grasp. Figure 10 and its accompanying explanation offer an example.

To prepare an earth for a harvest of grain requires a very special physical universe. To bake it into bread requires the intellectual will to do so. Had the gold rim of nature signaled a deterministic world, the concept of will would be meaningless. Quantum mechanics has revealed that within the boundaries of nature there is a given leeway that provides the space for choice and the will to choose.

"In this explanation lies the secret of the Table [and its gold rim]. For since the time that the world came into existence, God's blessing did not create something from nothing; however, the world follows its natural course."[17]

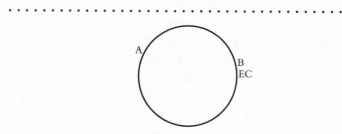

FIGURE 10: Human beings cannot envision a world in which there are more than the four dimensions we can sense physically. To illustrate the impact additional "invisible" dimensions might have, I offer this fantasy:

We are circle people. We live as dots on a circle, and we know only the line of the circle as we glide along this line forward and backward. For us there is no up or down, right or left, or center to the circle. Since the center doesn't lie on the line of the circle, we can't even conceive of a center.

Dot A and dot B live 20 miles apart on the line of the circle. Over the years B has also become friendly with EC, who moved into the area from another location. In circle world, the speed at which light travels (the fastest speed possible) is 10 miles per hour. B's motorcycle can reach that speed, so B can visit A in two hours. One day B receives a message from A to come quickly: A has met dot D, the perfect dot-mate for B. As B zooms off at the speed of light, he yells over his shoulder to EC the good news of his impending marriage. Two hours later, B screeches to a stop at A's home, only to find EC receiving congratulations from A on *his* marriage to D—over whom he is simply dotty.

But isn't this impossible? If B traveled to A at the speed of light, there was no way that EC could have gotten there sooner. Of course for EC there was a way: Unknown to A and B, EC is extracircular, and for him to reach A before B, he merely traveled in a straight line across the center of the circle, an act incomprehensible to A, B, and D because for dot people there is no such thing as a center.

There are a host of replicable scientific experiments whose results cannot be explained according to a four-dimensional worldview. Five or more dimensions would satisfy the requirements. Perhaps those exist but in a realm not comprehended by our senses. B saw EC married to D, but couldn't explain EC's speedy arrival, just as we can observe the results of these confusing experiments, though we can't explain the causes. Biblically, the fifth dimension might be a unifying spirituality.

Well, What About Dinosaurs?

I've managed to skirt the issue for most of this book. Now the moment of truth has come. What about dinosaurs? Why doesn't the Bible mention dinosaurs? Of course the Bible has no obligation to mention them. It doesn't mention bananas or oranges either. No one seems to have a problem with any of the other omissions, just with dinosaurs. There's a logic to the question. Dinosaurs represent a dramatic, tangible, even newsworthy fact by which science seems to challenge the Bible as an ultimate source of truth. Fortunately the Bible does mention dinosaurs, though not by that name. Dinosaur is a fairly modern word, a composite of two Greek words meaning terrible lizard. That is a close approximation of the Bible's description of these beasts.

Though I'd been asked about dinosaurs many times, still I never expected to hear the question in China, and certainly not three miles into the lush farmland of Jiangsu province, a locale that serves as the region's bread and tea basket.

Indeed food, or the lack of it, had brought me back to China. Using my background in nuclear physics, I'd stumbled onto a way of significantly increasing fish production in the tropical and semitropical ponds that dot the landscape of southeast Asia. It was and still is a good feeling to see the extra portions reach the white enameled bowls of the local populace.

Transferring that technology to third world farmers had become for me a personal mission. And Jiangsu was the ideal location to accomplish that task. It is home to the International Fisheries Research and Training Center—China's first such facility since the cultural revolution collapsed. In 1981, under United Nations auspices, D. H. Buck and I had helped design the research center, located three miles west of the city of Wuxi. Now I had returned to direct the scientific programs.

From the lab buildings a narrow, three-mile trail pushes its way through the aggressive growth of semitropical vegetation to an array of hand-dug fish ponds. Only at the crest of the last hill before the ponds do the branches give way, opening to a spectacular view of mountains washed in a gouache of white by the ever present mist.

It may have been the rush of emotion aroused by that view coupled with the guaranteed privacy of the location that prompted my Chinese colleague to risk a question about religion. He certainly had never even hinted at the subject before. Chinese communism is so purely a materialistic culture that religion seemed to have no place. For all I knew it may have been illegal as well.

We had stopped to catch our breath and admire the view. Without facing me he said "My grandmother has become a churchgoer. She never was before. She told me she believes the Bible is true."

If he expected a quick reply, I disappointed him. I was not about to put my freedom in jeopardy by touting biblical religion in this secular country. Two days earlier I had received a reprimand when I left a group of Chinese scientists to photograph a Pont-Aven-like scene of workers loading farm produce onto a small barge. "Remember you're a foreigner here. We expect you to keep a low profile." I understood the warning.

Now my colleague from Wuxi was probing for answers. But with his question, I knew his was not a government-inspired ploy to test me. "If the Bible is true, why doesn't it teach about dinosaurs?"

Dinosaurs—even here. I had come to Wuxi to integrate physics with farming. Zeng Li was asking me to integrate science with religion.

In retrospect, I realize I might have anticipated his question. The extensive find of dinosaur fossils in the Gobi desert had caught the imagination of the Chinese press. For Zeng Li they were one more hitch in his struggle to reconcile the secular culture of which he was a part with the spiritual needs his grandmother had awakened in him.

It caught Zeng Li by surprise when I showed him where dinosaurs are mentioned in the Bible. In Genesis 1:21 we are told that on day five God created the basis for all animal life. Among the categories of animals listed is one named *taninim gedolim. Gedolim* means big, and so we read "the big *taninim.*" Pick up five different English translations of the Hebrew Bible and you're likely to find five different meanings for the word *taninim:* whales, alligators, sea monsters, even dragons. Yet *taneen,* the singular of *taninim,* is a word that appears elsewhere in the Bible and its meaning is known.

In Exodus 3, the Eternal spoke to Moses from the burning bush and told him to return to Egypt to lead the enslaved Hebrews to freedom. Moses felt incapable of the task and so the Eternal gave him several signs, one related to his shepherd's staff. When Moses was told to throw his staff on the ground "it became a *nahash*" (Ex. 4:3). *Nahash* is the Hebrew word for snake. After Moses' return to Egypt, when Pharaoh asked for a sign, Moses' staff was again thrown to the ground and "became a *taneen*" (Ex. 7:10). Why didn't it become a *nahash,* a snake? And just five verses later, the Eternal tells Moses: "Get to Pharaoh in the morning, behold he goes to the water, and stand by the river's edge and the staff which turned into a *nahash* take in your hand" (Ex. 7:15).

It's the same staff. The change is first referred to as a *nahash,* then as a *taneen,* then as a *nahash.* We know that *nahash* means snake from its use elsewhere. *Taneen* must be a general category of animals since it appears in the creation chapter of Genesis where, other than Adam, only general categories of life are listed. So *taneen* must be the general category within which *nahash*—snake—falls. The general category for snakes is reptile. Thus Genesis 1:21 translates as: "And God created the big reptiles . . ." The biggest reptiles were the dinosaurs. But the author of Genesis did not specify dinosaurs directly, because that would have been inconsistent with the pattern of the chapter. The entire account of Genesis is stated in terms of objects known or knowable to the myriad of witnesses present at Sinai 3,300 years ago. Dinosaurs were not part of their world. But the hint, along with so many other hints, was there in the text for later generations to discover.

That discussion about dinosaurs was the first of many I had with Zeng Li. Over the years, in the privacy of that ridge before the ponds, we spent hours resolving supposed conflicts between science and the Bible. Years have passed since my last visit, and Zeng Li now joins his grandmother at religious services when research time permits. But for many persons the debate that underlies the question of dinosaurs remains.

In science we search for the "how" of the universe. We study the world and hope to learn the laws by which it functions. But at the end of the day, the questions we all ask relate not to the how but to the why of existence. Is there meaning to our lives that transcends the splendor we see about us? Something that goes beyond the physical? And if so, how can we probe that meaning? So much of life seems to hang upon uncontrollable events. Was it chance or was it teleology that killed the dinosaurs? And what of the meteor that almost killed us all?

March 23, 1989. There was no panic in the streets, no race for deep bomb shelters. Life went on as usual. Spring training was in full swing.

No astronomer saw it coming. A week would pass before anyone discovered what had—or better said—what had not happened. An asteroid was heading toward Earth. Not a giant the size of which killed off the dinosaurs. This more modest cousin was a mere kilometer in diameter. It represented enough energy to destroy most of the life on any continent it struck, something like the simultaneous explosion of twenty thousand one-megaton hydrogen bombs.

It whipped past us at 72,000 mph, crossing our exact path but missing by a mere six hours. For five billion years that rock, an agglomeration of a billion tons of star dust, had circled in the solar system just as had Earth. Of those five billion years, it missed us by six hours. That is like tuning an experiment to an error margin of one part in seven million million, a precision rarely reached in the laboratory.

January 1950. By the time the train reached our stop on rural Long Island, the receding storm had deposited another ten inches of snow. That brought the total to three feet during the previous two weeks. The trip from Penn Station had a festive air. College students home for winter vacation mixed with suburbanites finishing a day in Manhattan. The holiday lights that decorated homes along our route played on the crystal frost lining the train's windows. Steam hissed from the heaters that edged the floors.

"Hicksville. Change here for the Ronkonkoma line. It's slippery so watch your step getting offooppps!" The conductors loved to joke. We bundled up scarfs, coats, and gloves and headed to the railroad parking lot.

The high square silhouette of our car was easy to spot. The vintage four-door 1928 Buick was replete with wood spoke wheels, velvet-lined interior sporting built-in flower vases, and a set of levers on the steering wheel that allowed the driver to control everything from the air-fuel mixture to timing the spark. Several fellow travelers spun their wheels before giving up to trudge home.

The ten inches of new snow were no match for the Buick's foot-and-half-a-high chassis. We offered rides to three of the walkers. The only track was the one left by us. Three-foot snow walls gave clues to where the road led. In those days the four-mile drive to Brookville encountered only one traffic light, that as we crossed Jericho Turnpike. Ten years would pass before vegetable and potato fields yielded to housing developers and the expressways that now criss-cross the Island.

We parked the car among the roadside trees in a space Dad and I had shoveled that morning before heading off to New York. There was no chance of driving up to the house. The hundred-meter-long driveway that connected road to home by winding its way through our apple orchard was completely snowed in. And in any case, the walk tonight would be a pleasure. The storm had left the air crystal clear, stars like blazing pinholes in the moonless velvet sky. The frozen snow glittered in their silent twinkling light.

And then, what a scare! Shadows shot across the suddenly illuminated snow. Not the kind that in good behavior keep their logical direction. These shadows in a spit of time swept a circle of deep purple that ran around our legs, mirroring in negative the blue-white flash a huge meteor had cut across the sky. Ten-thousand-degree heat vaporized the white-hot rock. A glowing train of ionized air marked its path. For five billion years that stone had circled the sun in the vacuum of space. Its temperature had been the black of the night sky, 486 Fahrenheit degrees below the freezing point of water. Just 5 Fahrenheit degrees above absolute zero, a temperature close to the heat death of the universe. Now that piece of star-stuff was pushing its way through the top of our atmosphere at 70,000 miles an hour. At that speed, even the rarefied gases of the stratosphere produce an enormous frictional rub. The flash of light in which we stood gave its bright testimony to those forces.

That meteor might have glanced down instead of up as it hit Earth's atmosphere. In place of exquisite beauty, disaster would have struck. Rare though such events may be, they can give the feeling that life is a game of chance. Make your plans for the future and hope the cosmic rules won't change. If they do, you can be washed out in a flash.

Life as a dice game is one conclusion. But remember, the mammals made it through the disaster 65 million years ago. It was the dinosaurs that lost the race. Perhaps, instead of fortuity, we have discovered a cosmic tuning to the flow of life. Rather than chance, the fossils that mark the demise of the dinosaurs may be evidence for a teleology.

Plato likened our perception of life to persons viewing shadows on a wall while unaware of the far grander reality that produced those two-dimensional images. Three hundred years earlier, the prophet Isaiah had laid the basis for Plato's analogy:

> *Then the eyes of the blind shall be opened and the ears of the deaf unstopped. . . . The people that walked in darkness have seen a great light; they that dwelled in the land of images, upon them the light has shone. (Isaiah 35:5, 9:1)*

A. COSMIC BACKGROUND RADIATION (CBR)
AS A UNIVERSAL CLOCK

The cosmic clock of Genesis is based on the characteristics of CBR (wavelength, frequency, temperature). An observer can record these characteristics only as she or he receives the radiation. Owing to the expansion of the universe and the accompanying stretching of space, these characteristics may change considerably from what they were when the CBR was first emitted. If received frequency (the number of wave cycles per second) is taken to represent the rate at which the Genesis clock "ticks," then increasing wavelength and lower wave frequency, caused by the stretching of space, indicate a slowing of the Genesis clock and therefore a change in the perception of time when viewing events that occurred in the distant past.

I offer the following example to help clarify this principle of altered perception of time. Imagine that during the days of mail delivery via pony express, you receive a letter from a friend and in it he promises to send you another letter in six days. Unbeknownst to you, during those ensuing six days he has been traveling away from you. As promised, on day six he sends off the letter. But now, because of his travels, delivery takes an additional six days and so the letter arrives twelve days after the previous one. If the interval between the letters marked the passage of time, the receiver of the letters would assume that twelve of his days equaled six of the sender's days. The perception would be that the sender's time was passing half as fast as the receiver's.

In the letter the sender (S) wrote that he would now send a letter each day. Unbeknownst to the receiver (R), S now travels back toward R with the pony express mail. Each day as promised, S sends off a letter. Because S is now traveling back toward R, each letter takes a day less than the previous letter to arrive, which means that the six letters all arrive one right after the other, only minutes apart.

R can only assume that the time of S during the past six days was going much faster than his. The perception by R is that a full day has passed for S in a few minutes—the time between the arrival of each letter. If the letters were and are the only contact R and S ever have, it will never be possible to know whether the changed rate of time's flow truly happened or was only perception. We know only because we see both R and S, something which R and S cannot do.

Let's relate this idea to the changing CBR. In place of the traveler's letters we'll have an emitter's radiation, and in place of the receiver of letters we'll have an observer of the radiation. Since radiation frequency and radiation temperature are directly related, I'll refer to temperature as if it were frequency. Assume CBR is emitted at 3×10^{12}°K (the CBR temperature at about the time quarks confined into protons and neutrons). Traveling through time and space, the CBR is observed much later at 3°K (the approximate current CBR temperature). The emitter informs us that he sent 3×10^{12} waves each second, but the observer receives only 3 waves each second, or a million million times fewer than claimed. A million million seconds of observation are required to receive the information that the emitter claims to have sent in one second, and six million million days for the information sent in six days. Six million million days equals 16 million years! The effect of the expansion of the universe on radiation is that the same expansion factor applies to observe rates of events in the past. (See chapter 3, notes 13 and 14.)

In fact the passage of local proper time may be the same in both eras, but due to the stretching of space as the information passed between them (known as the redshift when referring to radiation frequency), the perception of relative time is very different.

In this manner, the days of Genesis, as viewed from the start of day one, might indeed remain days and yet contain all the bygone ages of the universe, as we view those ages looking from the present back through the distant reaches of space and time.

B. PROBLEMS IN ESTIMATING THE AGE OF THE UNIVERSE

Discussions of the age of the universe and divergent scientific opinions on that age appear frequently in the popular press. Therefore a basic understanding of how the age is calculated and of the problems encountered is in order.

One method for estimating the age of the universe is based on the assumption that if we can calculate how fast the universe is expanding, we can extrapolate that expansion back in time to the moment when the entire universe was contained within a minuscule speck of space, that is, the instant of the big bang, the creation. Extrapolation is the key to all aspects of this calculation.

To measure the rate of expansion, we must measure the rate at which distant galaxies are moving away from our galaxy. These galaxies must be sufficiently far from us so that the mutual gravitational attraction between them and our galaxy, the Milky Way, is small. Otherwise our Milky Way will in a sense be attached by gravity to the galaxy being measured, and the motions will not be independent. Therein lies the problem. The further a galaxy is from us, the less the mutual gravitational attraction, the more independent the motion, but the less accurate the measurement of the distance to that galaxy and hence the less accurate the estimate of the rate of expansion (the Hubble constant).

Direct measurement of distances to stars by trigonometric methods (measuring the change in the angle to the chosen star as the Earth proceeds around the Sun) is limited to stars within a distance of several hundred light years. (One light year equals 9.5 trillion [9.5 x 10^{12}] kilometers.) For more distant stars, a favored method for estimating distance is related to the pulsing of Cepheid variable stars. These stars pulse at fixed rates from once per several minutes to once per several months.

The pulse rate for Cepheids in the Milky Way appears to be related to their absolute brightness. If this relationship between brightness and pulse rate is true for the entire universe, then a measure of the pulse rate of a distant Cepheid is a measure of its absolute brightness. Comparing its apparent brightness (how bright it appears to us) to its absolute brightness (a value extrapolated from its pulse rate) gives an estimate of that Cepheid's distance from Earth. Estimating the distance of a hundred-watt light bulb from its observed brightness would be a parallel example. We know how bright the bulb shines when it is close. The dimmer it appears to be, the farther away we know it is.

A measure of the Doppler stretching of the light waves from the galaxy in which the Cepheid is located (known as the red shift) gives a measure of how fast that Cepheid's galaxy is moving away from the Milky Way, provided the red shift is due totally to the galaxy's motion (gravity also stretches light waves). From there it is one step to calculating the Hubble constant and from that the age of the universe.

Notice how many assumptions are made. All Cepheids are assumed to have the same ratio of pulse rate to absolute brightness. We have only shown this for stars in our immediate vicinity. Yet the entire fabric of this age estimate rests on this assumption holding for the entire universe. We can only make the Cepheid brightness-to-distance calculation out to well under 100 million light years, but the universe extends to some 10 to 20 billion light years. We are measuring much less than 1 percent of the universe and assuming it represents the entire universe.

To get a feel for the magnitude of this extrapolation, on a piece of paper draw a circle one inch in radius. We are at the center. Two or three points on that circle represent the "measured" universe. Now, keeping the same center, draw another circle 14 feet in diameter. The larger circle represents the entire universe. Compare the sizes of the two circles. Clearly, we have measured very little of the universe. And all cosmological data tell us the universe is not uniform.

In addition, we have assumed that expansion has been constant or predictable over the billions of years since the big bang. We know (as well as we can ever "know" facts about the first fractions of a second of the universe) that the universe initially expanded at a rate billions of times faster than it is expanding today. At what time the expansion settled into a predictable pattern is speculation, and yet this crucially affects the age estimation.

An alternative method of age estimation is based on the elemental composition of stars and how the elements formed after the big bang. This too requires a series of assumptions. The oldest stars (first-generation stars) formed from the primordial hydrogen and helium produced shortly after the big bang. These first-generation stars are poor in the heavier elements since those elements were made later in the explosions of other stars (supernovae). Younger stars (second- and third-generation stars, such as the Sun) include these heavier elements since they formed from a mixture of primordial hydrogen and helium plus the residue (star dust) of supernovae. The chemical composition of the first-generation stars in conjunction with their sizes provides the basis by which their ages are estimated. In the Milky Way, these stars are found in globular clusters outside of the Milky Way's spiral, the spiral being younger than the outlying clusters.

Considering the extent of the assumptions involved and the complete technical independence of these two methods, it is encouraging that the estimates fall within a factor of two, the range being ten to twenty billion years.

The million-million-to-one ratio between Genesis time during the six days of creation and our current perception of time is not an extrapolation. That ratio is based on the known threshold temperature at which matter (protons and neutrons) forms, a value measured in today's physics laboratories, and the measured temperature of the black of space, the cosmic radiation background.

C. THE LOGIC OF HAVING A BIBLICAL CALENDAR

The biblical calendar is derived in part by summing the ages of persons mentioned in the Bible. If the ages were not included, then there would not be a problem with a ten- to twenty-billion-year age for the universe or million-year-old fossils. There would not be any biblical ages with which to compare them. Clearly, the Bible includes the ages for a purpose.

A calendar might not seem crucial for the social or theological goals of the Bible. That the city of Ur was a flourishing center of idol worship some 3,900 years ago (Gen. 11), or that Joshua fought the battle of Jericho 3,400 years ago, in the spring, at the harvest (Josh. 3 and 6) could be superfluous knowledge in a text aiming to develop a just and moral society. Apparently the author felt the peripheral information was essential toward the overall goal. The prophet Isaiah explained it as follows: "The Lord was pleased for His righteousness to enlarge and glorify the Torah" (Is. 42:21).

This enlarging of the Torah refers to the "extra" information such as that which leads to a calendar.[1] While there is no way of *proving* that it is a benefit to "love your neighbor as yourself" (Lev. 19:18), or to observe the sabbath (Deut. 5:12), we can prove (through archaeology and related sciences) that the Bible's descriptions of Ur and Joshua and the many other biblical accounts of ancient events are accurate. If these provable claims are true, then perhaps those biblical claims that can't be proven, such as the benefits for respecting a neighbor, not bribing, not cheating in business, are also true. After all, they all are found in the same Book.

D. THE LONG LIFE SPANS AT THE TIME OF ADAM AND EVE

A question arises when we confront the very long ages of pre-Noah biblical personages. We are told Methusalah lived 969 years (Gen. 5:20). Jered lived 962 years (Gen. 5:20). During the ten generations from Adam to Noah, ages of 600 years were the norm. Is this possible?

The time represented by these generations must have passed. This we know from the fact that biblical dating of events for the period between Adam and Noah matches the archaeological dates for the corresponding events. If the years had not passed, the dates would not match. Some claim that each person is actually a combination of several generations. I propose that these ages were actually reached.

Aging is a metabolic process. Each species is programmed for death at a given age range.[2-4] Hence fleas live to five years, dogs to about fifteen years, and humans about seventy years. Fleas never reach the age of seventy. Their genetic package does not allow them to.

There are terrible mutations that upset the delicate aging process. Progeria speeds up the aging process almost tenfold, causing a teenager to die with the body of an old person. Within the realm of possibilities is the reverse process, slowing aging tenfold. It would be surprising but not inconceivable that manipulation of a flea's genome might allow it to live ten times longer than normal, thus reaching the age of fifty years. After all, several animal species live even longer than fifty years. The fact that no animals currently reach the long ages associated with pre-Noah biblical persons does not preclude the possibility that this potential exists within our genome.

If human metabolism was slower and life spans were longer during the pre-Noah period, fossils would not indicate this. The slower metabolisms would result in fossils that appear to have formed from younger individuals.

The range of ages for first giving birth and for deaths in those pre-Flood generations indicates that a variety of metabolic processes was extant. According to population genetics, the environment selects from that variety the most advantageous aspect of the trait. Nahmanides suggests that the shortening of life spans after Noah's time and following the population dispersion after the destruction of the Tower of Babel (Gen. 11:1ff) resulted from a more demanding climate.[5] Maimonides relates it to changes in diet. Both suggestions sound like environmental selection. Whatever the cause, by the time of Abraham, ten generations after Noah, Sarah's giving birth at ninety years of age required an explicit miracle (Gen. 18:10ff).

Comparison of the ages of the first giving birth and of death for persons living before Noah with the corresponding ages today reveals an instructive relationship. In pre-Noah times, the ages at the birth of a firstborn were 100 to 130 years. Age at death ranged approximately from 600 to 900 years. Today, puberty is usually reached between eleven and thirteen years of age. Death comes at sixty to ninety. Note the relationship. The pre-Noah ages for both puberty and death are tenfold higher than today, as if the entire process was slowed by a factor of ten. This could have been caused by a metabolic shift.

These long ages give rise to an interesting speculation discussed fifteen hundred years ago in the Talmud.[6] Psalms states that a thousand generations were to pass before the Torah was given: "He remembers forever His covenant, the matter commanded to the thousandth generation" (Ps. 105:8). But, the Talmud notes, only twenty-six generations passed between Adam receiving the soul of humanity, the *neshama,* and Moses receiving the Torah at Sinai. Where are the missing 974 generations? The Talmud presents several suggestions. The following is my conjecture.

If the average span of a generation was 130 years (pre-Noah data), those 974 generations would have lasted some 130,000 years. Were the Neanderthal and Cro-Magnon those missing generations that were never created? Recall that the creation relates to the creation of the *neshama* and not to the making of the body.

E. GENESIS DAY THREE

The Bible records the formation of the seas and the appearance of dry land on day three of the six days of creation. Then we are told that the Earth brought forth grass and herbs and fruit trees (Gen. 1:9–13). This raises several chronological problems. The calculated time of day three is from 3.8 billion years ago until 1.8 billion years ago. That timing for the appearance of liquid water and the first plant life on Earth matches the paleontological date of 3.8 billion years in the past for both events. However, the first plant life was much less complex, consisting of single-celled bacteria and algae, not at all the biblical list of grass, herbs, and fruit trees. Land plants appear in the fossil record only some 400 million years ago and flowering plants and trees some 120 million years ago. Can paleontology be reconciled with the biblical dates?

Nahmanides' seven-hundred-year-old commentary on Genesis 1:12 provides the information needed for establishing consonance between these two sources of knowledge. "There was no special day assigned for this command for [the appearance of the various forms of] vegetation alone, since it is not a unique work." Though the basis for advanced forms of plant life was set on day three, according to the Bible the actual plants did not necessarily appear then. Molecular biology has discovered that some forms of single-celled algae have as much as one hundred times the amount of DNA (genetic information) per cell as do mammals.[7] A genetic library that large could indeed contain the basic information for the forms of plant life that appeared much later.

The rationale for Nahmanides' commentary is found elsewhere in Genesis. When the Bible is relating a topic of immediate but not continuing interest, it condenses the chronology of that topic and presents the entire account in one place rather than break into the narrative at a later point.

At the close of Genesis 11, we learn that Terah, the father of Abraham, took his family from Ur (a city in Mesopotamia) and traveled toward Canaan (modern-day Israel). Partway there, they settled in the city of Haran. Terah was seventy years old when Abraham was born (Gen. 11:26). He died in Haran at the age of 205 (Gen. 11:32). The very next verse (Gen. 12:1) tells us that the Eternal called to Abraham and told him to go to Canaan, which he and his household did. We learn that "Abraham was seventy-five years old when he left Haran" (Gen. 12:4). The order of the sentences would imply that Terah died and then Abraham left. But a simple bit of arithmetic reveals that Terah lived another sixty years after Abraham's departure. Abraham and Sarah have their first child, Isaac, twenty-five years later (Gen. 21) and still Terah had a further thirty-five years of life, but Terah is never mentioned after the close of Genesis 11. Terah was important because he was of the lineage leading to Abraham, Isaac, and Jacob. But he was not of continuing significance to the narrative. Rather than break into the text later during the account of Isaac's life to tell us of his death, Genesis condenses the account of Terah's life but gives us the needed information (the various ages) to make the accounting ourselves, and to learn this crucial technique of biblical exegesis.

The Bible is interested in plants, but the main focus is the flow of animal life leading to humanity and then to the line that leads to Abraham. As such, the Bible wraps up the account of plant life in two verses (Gen. 11–12), and relies on our understanding of biblical interpretation (e.g., Nahmanides' commentary) for the actual sequence. Only for plant life does Nahmanides tell us that the chronology is condensed. The topics of Genesis 1 related to animal life are of continuing interest as leading to humans, and so of the entire six days and thirty-one verses, two full days (five and six) and twelve verses (Gen. 1:20–31) are devoted to this sequence.

F. THE FLOOD AT THE TIME OF NOAH

Perhaps the most recalcitrant of problems in finding accord between Bible and science is the Flood at the time of Noah (Gen. 6–8), approximately 4,100 years ago. All life was to perish. Only Noah, his family, and two or fourteen representatives from each animal species were to survive by riding out the storm in a specially prepared ark (Gen. 6:17–22; 7:2).

Though flood stories are common to many ancient cultures, lines of native American civilizations show no break at the time of the biblical Flood. I do not know of any traditional commentaries that totally satisfy this discrepancy between science and Bible. There are, however, some clues that may be relevant. Nahmanides notes that there may have been persons other than Noah and his family who survived the Flood (based on Gen. 6:4). The ark rested on a mountain in Ararat, a region believed (but not definitely known) to be in Armenia (Gen. 8:4). But as Nahmanides notes, there are many other mountains in the world higher than those of Armenia (on Gen. 8:5). The Mediterranean Sea may not have been affected according to a two-thousand-year-old tradition.[8] These sources indicate that the flood may have been local, not global.

Mesopotamia, the region where Babel and Ur and presumably Noah's home as well are located, is bounded on three sides by the Tigris and Euphrates rivers. It is largely a plain. A massive flood approximately five thousand years ago produced a three-meter-thick sediment deposit in the vicinity of Ur.[9] So a flood at the time of Noah is not without precedent.

A colleague, Peggy Ketz, noted an important variation in the Hebrew text. The first mention of the Eternal's plan to destroy animal life is: "And the Eternal said I will blot out man whom I have created from on the face of the earth *(adamah),* both man and beast and creeping thing and fowl of the heavens" (Gen. 6:7). For the entire Flood account, the destruction is related to destroying life from the face of the *aretz,* not the *adamah.* Though *adamah* and *aretz* may both be translated as earth, they can also mean local environs and not the entire Earth. For example: "And the famine was on all the face of the *aretz*" (Gen. 41:56); and "There was no bread in all the *aretz*" (Gen. 47:13). Cain was banished "from the face of the *adamah*" (Gen. 4:14). He neither went to sea nor left Earth for Mars! We see that these terms, *aretz* and *adamah,* often have implications of limited geological extent.

The change in terminology from *adamah* to *aretz* may indicate a change in divine intent, a change from destroying all life to destroying life in the corrupt region of Mesopotamia. That change in intent is perhaps signaled by the verse following Genesis 6:7 that was quoted above: "But Noah found grace in the eyes if the Eternal" (Gen. 6:8). Things were bad, but not hopeless. Some good was found in the area. From this verse onward, destruction is limited to the face of the *aretz.*

References to the Bible and the commentaries herein are based on the following editions:

The Bible: The Holy Scriptures. Jerusalem: Koren Publishers, 1969. (Hebrew and English.)

The Babylonian Talmud. Transl. from the original Hebrew and Aramaic into English. London: Soncino Press, 1977.

Maimonides. *Guide for the Perplexed*. Transl. into English by M. Friedlander. London: G. Routledge & Sons, 1928.

Nahmanides. *Commentary on the Torah*. Ed. C. Chavel. Jerusalem: Rav Kook Institute, 1958 (Hebrew), 1971 (English trans.).

The Pentateuch (the Torah, or the Five Books of Moses), with commentary by Rashi in the original Hebrew, transl. into English. Jerusalem: Silbermann Family Publishers, 1973.

PROLOGUE

1.   D. Lapin, private communication.

CHAPTER I • HAS SCIENCE REPLACED THE BIBLE?

1.   E. Ladd, "Religion and American Values," *Society* 24:63–68, 1987.

2.   R. Feynman, R. Leighton, and M. Sands, *The Feynman Lectures on Physics,* vol. 1, Addison-Wesley, New York, 1963.

3.   Nahmanides, commentary on Exodus 25:24, 1250 C.E.

4.   S. Weinberg, "Life in the Universe," *Scientific American,* October 1994.

5.   M. Behe, *Darwin's Black Box,* Free Press, New York, 1996.

6.   T. H. Huxley, "On the Origin of Species," *Westminister Review,* April 1860.

7.   B. Rensberger, "Recent Studies Spark Revolution in Interpretation of Evolution," *New York Times,* 4 November 1980, C3.

8.   Babylonian Talmud Hagigah 11b, 12a, 500 C.E.

9.   J. M. Smith, *The Theory of Evolution,* Penguin Books, New York, 1958.

10.   S. J. Gould, *Wonderful Life,* W. W. Norton, New York, 1989.

11.   J. Levington, "The Big Bang of Animal Evolution," *Scientific American,* November 1992.

12.   D. Erwin, "The Mother of Mass Extinctions," *Scientific American,* July 1996.

13.   J. Marx, "DNA Replication," *Science* 270:1585, 1995.

14. J. Hertz, Commentary on Deut. 30:9, *The Pentateuch and Haftorahs,* Soncino Press, London, 1963.

15. Maimonides, *Guide for the Perplexed,* Introduction and part 3:51, 1190 C.E.

16. J. Horgan, "Heart of the Matter," *Scientific American,* December 1993.

17. R. Kerr, "Did Darwin Get It All Right?," *Science* 267:1421, 1995.

18. J. Grotzinger et al., "Biostratigraphic and Geochronological Constraints on Early Animal Evolution," *Science* 270:598–604, 1995.

19. S. J. Gould, "Impeaching a Self-Appointed Judge," *Scientific American,* July 1992.

20. P. Johnson, *History of the Jews,* Harper & Row, New York: 1987.

CHAPTER 2 • THE NEW CONVERGENCE

1. R. Penrose, *The Emperor's New Mind,* Penguin Books, New York, 1991.

2. P. Davies, *The Mind of God,* Simon & Schuster, New York, 1992.

3. S. Weinberg, *The First Three Minutes,* Basic Books, New York, 1977.

4. S. Brush, "How Cosmology Became a Science," *Scientific American,* August 1992.

5. J. Maddox, "Down with the Big Bang," *Nature* 340:425, 1989.

6. H. Pagels, *Perfect Symmetry,* Michael Joseph, London, 1985.

7. J. Maddox, op. cit.

8. Babylonian Talmud Hagigah 12A, ca. 500.

9. Nahmanides, Introduction to *Commentary on Genesis,* ca. 1250.

10. E. Tryon, "Is the Universe a Vacuum Fluctuation?" *Nature* 246:396, 1973.

11. D. Atkatz, "Quantum Cosmology for Pedestrians," *American Journal of Physics* 62:619, 1994.

12. L. Orgel, "The Origin of Life on the Earth," *Scientific American,* October 1994.

13. J. Rebek, Jr., "Synthetic Self-Replicating Molecules," *Scientific American,* July 1994.

14. C. DeDuve, *Blueprint for a Cell,* Neil Patterson Publishers, Burlington, NC, 1990.

15. J. Levinton, "The Big Bang of Animal Evolution," *Scientific American,* November 1992.

16. S. Gould, "The Evolution of Life on Earth," *Scientific American,* October 1994.

17. M. Nash, "When Life Exploded," *Time,* 4 December 1995.

18. R. Kerr, "Timing Evolution's Early Bursts," *Science* 267:30, 1995.

19. J. Kaiser, "A New Theory of Insect Wing Origins," *Science* 266:363, 1994.

20. J. Gillespie, "Rates of Molecular Evolution," *Annual Review of Ecol. Syst.* 17:637, 1986.

21. M. Denton, *Evolution: A Theory in Crisis,* Burnett Books, London, 1985.

22. R. Quiring et al., "Homology of the Eyeless Gene in Drosophila to the Small Eye Gene in Mice and Aniridia in Humans," *Science* 265:785, 1994.

23. J. Fischman, "Why Mammal Ears Went on the Move," *Science* 270:1436, 1995.

24. D. York, "The Earliest History of the Earth," *Scientific American,* January 1993.

25. I. Dalziel, "Earth Before Pangea," *Scientific American,* January 1995.
26. J. Grotzinger et al., "Biostratigraphic and Geochronologic Constraints on Early Animal Evolution," *Science* 270:598–604, 1995.
27. D. Palmer, "Ediacarans in Deep Water," *Nature* 379:114–115, 1996.
28. J. M. Smith, *The Theory of Evolution,* 2nd ed., Penguin Books, New York 1975, p. 311.
29. C. Folsome, *Life: Origin and Evolution,* Scientific American special publication, 1979.
30. J. Horgan, "Trends in Evolution," *Scientific American,* February 1991.
31. J. Horgan, "Profile: Francis H. C. Crick," *Scientific American,* February 1992.
32. S. J. Gould, *Eight Little Piggies,* W.W. Norton, New York, 1993, pp. 225, 226.
33. S. J. Gould, *Bully for Brontosaurus,* Penguin Books, New York, 1992, p. 39.
34. Gould, *Eight Little Piggies,* pp. 79, 179, 217.
35. J. Wilford, "Spectacular Fossils Record Early Riot of Creation," *New York Times,* 23 April 1991, C-1.
36. Nash, op. cit.
37. M. Behe, *Darwin's Black Box,* Free Press, New York, 1996.

CHAPTER 3 · THE AGE OF OUR UNIVERSE

1. M. Bolte and J. Hogan, "Conflict over the Age of the Universe," *Nature* 376:399–402, 1995.
2. W. Freedman et al., "Distance to Virgo Cluster Galaxy M100 from Hubble Space Telescope," *Nature* 371:757–762, 1994.
3. Rashi, in his commentary on the 1,500-year-old Talmud Hagigah 12A, referring to " . . . day one" (Gen. 1:5).
4. Nahmanides, commentary on Genesis 1:3, 1250. C.E.
5. Nahmanides on Exodus 21:2 and Leviticus 25:2.
6. Maimonides, *Guide for the Perplexed,* Introduction and part 3:51, 1190 C.E.
7. Leviticus Rabba 29:1, 300 C.E.
8. Jerusalem Talmud Avodah Zarah 1:2, 350 C.E.
9. Babylonian Talmud Hagigah 13B, 14A. year 500 C.E.
10. See notes 7–9.
11. C. Misner, K. Thorne, and J. Wheeler, *Gravitation,* W. H. Freeman, San Francisco, 1971, pp. 659, 776.
12. S. Weinberg, *Gravitation and Cosmology,* John Wiley & Sons, New York, 1972, pp. 80, 540.
13. M. Fukugita, C. Hogan, P.J.E. Peebles, "The History of the Galaxies," *Nature* 381:489, 1996.
14. P.J.E. Peebles, *Principles of Physical Cosmology,* Princeton University Press, Princeton, 1993, pp. 71, 91, 96, 135.
15. J. Wilford, "Believers Score in Battle over the Battle of Jericho," *New York Times,* February 1990, A8, 22.
16. Misner et al., op. cit.
17. J. Levy-Leblond, "The Unbegun Big Bang," *Nature* 342:23, 1989.
18. J. Levy-Leblond, "Did the Big Bang Begin?" *American Journal of Physics* 58:156–159, 1990.

19. A. Grunbaum, "Pseudo-Creation of the Big Bang," *Nature* 344:821–822, 1990.

20. Fukugita et al., op. cit.

21. M. Harwit, "Cosmic Curvature and Condensation," *The Astrophysical Journal* 392:394–402, 1992.

22. Peebles, op. cit., pp. 91,96, emphasis added.

23. S. Weinberg, *The First Three Minutes,* Basic Books, New York, 1977.

24. S. Hawking, *A Brief History of Time,* Bantam Books, New York, 1988, p. 76.

25. P.J.E. Peebles, D. Schramm, E. Turner, and R. Kron, "The Evolution of the Universe," *Scientific American,* October 1994.

26. Peebles, op. cit., p. 96.

27. Nahmanides, commentary on Gen. 1:1, 4.

28. S. Weinberg, "Life in the Universe," *Scientific American,* October 1994.

29. J. Silk, *The Big Bang,* W.H. Freeman, New York, 1989, p. 72.

30. Talmud Hagigah 12A.

31. Nahmanides, commentary on Gen. 1:1, 2

32. Peebles, op. cit., p. 96.

33. Misner et al., op. cit.

34. Weinberg, "Life in the Universe," p. 1972.

35. Silk, op. cit.

36. B. Smith, "A Short History of the Universe," *National Geographic,* January 1994.

CHAPTER 4 • THE SIX DAYS OF GENESIS

1. C. Misner, K. Thorne, and J. Wheeler, *Gravitation,* W.H. Freeman, San Francisco, 1971, pp. 659, 776.

2. S. Weinberg "Life in the Universe," *Scientific American,* October 1994.

3. J. Silk, *The Big Bang,* W.H. Freeman, New York, 1989, p. 72.

4. P.J.E. Peebles, *Principles of Physical Cosmology,* Princeton University Press, Princeton, NJ, 1993.

5. Misner et al., op. cit.

6. A. Finkbeiner, "Closing In on Cosmic Expansion," *Science* 270:1295, 1995.

7. M. Fukugita, C. Hogan, P.J.E. Peebles, "The History of the Galaxies," *Nature* 381:489, 1996.

8. E. Maor, "The Story of *e,*" *The Sciences,* July/August 1994, pp. 24–29.

9. Nahmanides, commentary on Gen. 1:5.

10. Nahmanides, commentary on Gen. 1:4.

11. S. Weinberg, *The First Three Minutes,* Basic Books, New York, 1977.

12. P.J.E. Peebles, D. Schramm, E. Turner, and R. Kron, "The Evolution of the Universe," *Scientific American,* October 1994.

13. M. Bolte and J. Hogan, "Conflict over the Age of the Universe," *Nature* 376:399–402, 1995.

14. W. Freedman et al., "Distance to the Virgo Cluster Galaxy M100 from Hubble Space Telescope Observations of Cephids," *Nature* 371:757–762, 1994.

15. J. Maddox, "Big Bang Not Yet Dead but in Decline," *Nature* 377:99, 1995.

16. P. Cloud, *Oasis in Space,* W.W. Norton, New York, 1988, p. 167.
17. Weinberg, "Life in the Universe."
18. Nahmanides, commentary on Gen. 1:12.
19. C. Allegre and S. Schneider, "The Evolution of the Earth," *Scientific American,* October 1994.
20. Talmud Hagigah 12A; Rashi.
21. S. Gould, "The Evolution of Life on Earth," *Scientific American,* October 1994.
22. J. Levinton, "The Big Bang of Animal Evolution," *Scientific American,* November 1992.
23. M. Nash, "When Life Exploded," *Time,* 4 December 1995.

CHAPTER 5 • THE NATURE OF GOD

1. J. Mitter, "When Quails Come Back to Alexandria," *New York Times,* 31 October 1984, C3.
2. Nahmanides, commentary on Deut. 32:7.
3. Nahmanides, commentary on Lev. 19:2 and Lev. 21:6.
4. K. Miller, "Life's Grand Design," *Technology Review,* February/March 1994.

CHAPTER 6 • LIFE

1. G. Wald, "The Origin of Life," *Scientific American,* August 1954.
2. C. Folsome, *Life: Origin and Evolution,* Scientific American Special Publication, 1979.
3. C. DeDuve, *Blueprint for a Cell: The Nature and Origin of Life,* Neil Patterson Publishers, Burlington, NC, 1991.
4. M. Waldrop, "Did Life Really Start Out in a RNA World?" *Science* 246:246, December 1989.
5. L. Orgel, "The Origin of Life on Earth, *Scientific American,* October 1994.
6. J. Horgan, "Profile: Francis H. C. Crick," *Scientific American,* February 1992.
7. R. Kerr, "Did Darwin Get It All Right?" *Science* 267:1421–1422, 1995.
8. J. Wilford, "Spectacular Fossils Record Early Riot of Creation," *New York Times,* 23 April 1991, p. C-1.
9. J. M. Smith, *The Theory of Evolution,* 2nd ed., Penguin Books, New York, 1975.
10. S. Bowring, J. Grotzinger, and C. Isachsen, "Calibrating Rates of Early Cambrian Evolution," *Science* 261:1293, 1993.
11. R. Gore, "The Cambrian Period Explosion of Life," *National Geographic* 184:125, October 1993.
12. J. Grotzinger et al., "Biostratigraphic and Geochronologic Constraints on Early Animal Evolution," *Science* 270:598, 1995.
13. D. Palmer, "Ediacarans in Deep Water," *Nature* 379:114, 1996.
14. R. Britter and D. Kohne, "Repeated Segments of DNA," *Scientific American,* April 1970.
15. B. Hall, "Development Mechanisms Underlying the Formation of Atavisms," *Biological Review* 59:89, 1984.

16. P. Gingerich et al., "Origin of Whales in Epicontinental Remnant Seas," *Science* 220:403, 1983.

17. A. Berta, "What Is a Whale?" *Science* 263:180, 1994.

18. R. Quiring, U. Kloter, and W. Gehring, "Homology of the Eyeless Gene in Drosophila to the Small Eye Gene in Mice and Aniridia in Humans," *Science* 265:785–789, 1994.

19. C. Zuker, "On the Evolution of eyes," *Science,* 265:742, 1994.

20. Smith, op. cit.

21. Quiring et al., op. cit.

22. J. Kaiser, "A New Theory of the Insect Wing Origins," *Science* 266:363, 1994.

23. J. Marden and M. Kramer, "Surface Skimming Stoneflies: A Possible Intermediate Stage in Flight Evolution," *Science* 266:427, 1994.

24. A. Feduccin, *Age of Fossils,* Harvard University Press, Cambridge, MA, 1980.

25. R. Gore, "Dinosaurs," *National Geographic* 183:2, January 1993.

26. P. Wellnhefer, "Archaeopteryx," *Scientific American,* May 1990.

27. Gore, op. cit.

28. Nahmanides, commentary on Genesis 1:5.

29. D. Nof and N. Paldor, "Are There Oceanographic Explanations for the Israelites' Crossing of the Red Sea?" *Bulletin of the American Meteorological Society* 73:305, 1992.

30. D. Nof and N. Paldor, "Statistics of Wind over the Red Sea with Application to the Exodus Question," *Journal of Applied Meteorology* 33:1017, 1994.

CHAPTER 7 · EVOLUTION

1. W. Gitt, "Information: the Third Fundamental Quantity," *Siemens Review* 56(6) November/December 1989.

2. R. May, "How Many Species Inhabit the Earth," *Scientific American,* October 1992.

3. J. Levinton, The Big Bang of Animal Evolution, *Scientific American,* November 1992.

4. S. Gould, "The Evolution of Life on the Earth," *Scientific American,* October 1994.

5. R. Gore, "The Cambrian Explosion of Life," *National Geographic,* p. 125, October 1993.

6. M. Nash, "When Life Exploded," *Time,* December 4, 1995.

7. R. Kerr, "Evolution's Big Bang Gets Even More Explosive," *Science* 261:1274, 1993.

8. R. Dawkins, *The Blind Watchmaker,* W.W. Norton, New York, 1985, p. 94.

9. N. Patel, "Developmental Evolution," *Science* 266:581, 1994.

10. R. Burington, *Mathematical Tables and Formulas,* Handbook Publishing Co., Sandusky, Ohio, 1955.

11. A. Gibbons, "The Mystery of Humanity's Missing Mutations," *Science* 267:35, 1995.

12. L. Mettler, T. Gregy, and H. Schaffer, *Population Genetics and Evolution,* Prentice Hall, Englewood Cliffs, NJ, 1988, p. 104.

13. M. Radman and R. Wagner, "The High Fidelity of DNA Duplication," *Scientific American,* August 1988.

14. Ibid.

15. S. Bowring et al., "Calibrating Rates of Early Cambrian Evolution," *Science* 261:1293, 1993.

16. D. Canfield and A. Teshe, "Late Proterozoic Rise in Atmospheric Oxygen Concentration," *Nature* 382:127, 1996.

17. R. Quiring et al., "Homology of the Eyeglass Gene in Drosophila to the Small Eye Gene in Mice and Aniridia in Humans," *Science* 265:785, 1994.

18. C. Zuker, "On the Evolution of Eyes," *Science* 265:742, 1994.

19. I. Prigogine, N. Gregoire, and A. Babloyantz, "Thermodynamics of Evolution," *Physics Today* 25:23, November 1972, and 25:38, December 1972.

20. L. Podolsky, private communication.

21. Gore, op. cit.

22. Kerr, op. cit.

23. P. Moorehead and M. Kaplan, "Mathematical Challenges to the Neo-Darwinian Interpretation of Evolution," *Proceedings of the Symposium, Wistar Institute of Biology,* 1967.

24. G. Halder et al., "Induction of Ectopic Eyes by Targeted Expression of the Eyeless Gene in Drosophilia," *Science* 267:1788, 1995.

25. R. Kerr, "Did Darwin Get It All Right?" *Science* 267:1421, 1995.

CHAPTER 8 • THE WATCHMAKER AND THE WATCH

1. R. Dawkins, *The Blind Watchmaker,* W.W. Norton, New York, 1986.

2. Talmud Keliim 8:5.

3. Maimonides, *Guide for the Perplexed,* 1:7

4. Ernst Mayr quoted in "The New Challenges," *Scientific American,* December 1992.

5. J. Rennie, "Darwin's Current Bulldog, a Profile of Ernst Mayr," *Scientific American,* August 1994.

6. A. Gibbons, "The Mystery of Humanity's Missing Mutations," *Science* 267:35, 1995.

7. Y. Coppens, "The Origin of Humankind," *Scientific American,* May 1994.

8. A. Kortlandt, "Rift over Origins," *Scientific American,* October 1994.

9. V. Morell, "African Origin: Westside Story," *Science* 270:1117, 1995.

10. F. de Wall, "Bonobo Sex and Society," *Scientific American,* March 1995.

11. R. Nowak, "Mining Treasures from Junk DNA," *Science* 263:608, 1994.

12. N. Davies and M. Brooke, "Coevolution of the Cuckoo and Its Hosts," *Scientific American,* January 1991.

13. M. Radman and R. Wagner, "The High Fidelity of DNA Duplication," *Scientific American,* August 1988.

14. Mettler et al., *Population Genetics and Evolution,* Prentice Hall, Englewood Cliffs, NJ, 1988.

15. Gibbons, op. cit.

16. Mettler, op. cit.

17. J. Rennie, "DNA's New Twists, *Scientific American,* March 1993.

18. C. Yoon, "The Wizard of Eyes: Evolution Creates Novelty by Varying the Same Old Tricks," *New York Times,* 1 November 1994, C1.

19. K. Lautwyler, "Chemical Guides Direct Young Neurons to Their Final Destinations," *Scientific American,* January 1995.

20. G. Halder et al., "Induction of Ectopic Eyes by Targeted Expression of the Eyeless Gene in Drosophila, *Science* 267:1788, 1995.

CHAPTER 9 • THE ORIGIN OF HUMANKIND

1. Rashi, commentary on Talmud Sanhedrin 97A 1050 C.E.; Leviticus Rabba 29:1.

2. J. Horgan, "The New Challenges," *Scientific American,* December 1992.

3. J. Rennie, "Darwin's Current Bulldog: Ernst Mayr," *Scientific American,* August 1994.

4. R. Penrose, *The Emperor's New Mind,* Penguin Books, New York, 1989.

5. T. Molleson, "The Eloquent Bones of Abu Hureyra," *Scientific American,* August 1994.

6. C. B. Walker, *Reading the Past: Cuneiform,* British Museum Publications, London, 1989.

7. B. Smith and W. Weng, *China: A History in Art,* Doubleday, New York, 1979.

8. Genesis Rabba 39:14.

9. Maimonides, Laws of idol worship.

10. Nahmanides, commentary on Gen. 2:7.

11. Talmud Keliim 8:5.

12. Maimonides, *Guide for the Perplexed;* Part One, Chapter 7.

13. Talmud Eruvim 18A.

14. Jerusalem Talmud Peah 1:1.

15. Nahmanides, commentary on Gen. 6:4.

16. Rashi, commentary on Gen. 3:6.

17. J. Healey, *The Early Alphabet,* British Museum Publications, London, 1990.

18. H. Frankfurt, "Freedom of Will and the Concept of a Person," in *Free Will,* ed. G. Watson, Oxford University Press, 1982.

CHAPTER 10 • THE SCIENCE OF FREE WILL

1. P. Yam, "Coming In from the Cold," *Scientific American,* August 1995.

2. W. Gitt, "Information: The Third Fundamental Quantity," *Siemens Review* 56(6):1–7, 1989.

3. J. Horgan, "Eugenics Revisited," *Scientific American,* June 1993.

4. S. LeVay, D. Hamer, and W. Byne, "Is Homosexuality Biologically Influenced," *Scientific American,* May 1994.

5. L. Wright, "Double Mystery," *The New Yorker,* 7 August 1995.

6. Talmud Chapters of the Fathers, 3:19.

7. Talmud Sanhedrin 90B.

8. Copied from the men's room wall of the Pecan Street Cafe, Austin, Texas, and quoted in E. Taylor and J. A. Wheeler's classic textbook *Spacetime Physics*, W. H. Freeman, San Francisco 1966.

9. Nahmanides, commentary on Ex. 3:13.

CHAPTER 11 • WHY BAD (AND GOOD) THINGS HAPPEN

1. *New York Observer*, 30 September 1991, p. 13.

2. M. Radman and R. Wagner, "The High Fidelity of DNA Duplication," *Scientific American*, August 1988.

3. P. Brou et al., "The Color of Things," *Scientific American*, September 1986.

4. H. Kushner, *When Bad Things Happen to Good People*, Avon Books, New York, 1983.

CHAPTER 12 • BREAD FROM EARTH

1. Nahmanides and Rashi, commentary on Ex. 25:24.

2. Talmud Avodah Zarah 54B

3. J. Horgan, "Heart of the Matter," *Scientific American*, December 1993.

4. S. Weinberg "Life in the Universe," *Scientific American*, October 1994.

5. B. Smith, "A Short History of the Universe," *National Geographic*, January 1994.

6. S. Hawking, *A Brief History of Time*, Bantam Books, New York, 1988.

7. S. Weinberg, *The First Three Minutes*, Basic Books, New York, 1977.

8. Weinberg, "Life in the Universe."

9. H. Pagels, *Perfect Symmetry*, pp. 277, 332, Michael Joseph, London, 1985.

10. P. Davies, private communication; S. Weinberg, private communication; A. Sakharov, as referenced in *Scientific American*, December 1993.

11. F. Press and R. Siever, *Earth*, W. H. Freeman, N.Y., 1986.

12. R. Penrose, *The Emperor's New Mind*, Penguin Books, New York, 1991.

13. Press and Siever, op. cit.

14. Penrose, op. cit.

15. S. Fetter et al., "Why Were Scud Casualties so Low?" *Nature* 361:293, 1993.

16. M. Behe, *Darwin's Black Box*, Free Press, New York, 1996.

17. Nahmanides and Rashi, op. cit. Emphasis added.

APPENDIX

1. Talmud Nida 51B, 500 C.E.

2. I. Zs-Nagy, D. Harman, and K. Kitani, "Pharmacology of Aging Processes," *Annals of the New York Academy of Science*, 717, 1994.

3. L. Margulis, "Sex, Death and Kefir," *Scientific American*, August 1994.

4. J. Rennie, "Immortal's Enzyme," *Scientific American*, July 1994.

5. Nahmanides, commentary on Genesis 5:4.

6. Talmud Hagigah 13B, 14A.

7. R. Britter and D. Kohne, "Repeated Segments of DNA," *Scientific American*, April 1970.

8. Genesis Rabba 32:19.

9. M. Dewsnap, "Uncovering the Deluge," *Biblical Archaeological Review* 22(4):56, July/August 1996.

ACKNOWLEDGMENTS

· ·

Intellectual endeavor, according to the Bible, comes in three forms: knowledge, understanding, and insight. The first comes directly from an individual source, such as a book. The second is attained by combining information from several sources to build something new. The third arises from an unknown source, inspiration—some might call it divine inspiration. I have built upon the help of many people, in some instances a single conversation, with others, many hours of discussion. Some of these people have advanced academic degrees. Some are ordained theologians. A few are Nobel laureates. I will only list the names here, without titles. I thank you all for helping to make *The Science of God* a reality.

Debra Harris and Beth Elon knew exactly where to direct the book. Bruce Nichols, my editor at The Free Press, made crucial suggestions in helping me realize the focus I was seeking. Edith Lewis answered my multitude of questions regarding style. Linnea Johnson's detailed copyediting improved the clarity of my writing. Helen Rees and Marc Jaffe years ago convinced me that many people were eager to learn about the unfolding confluence of modern scientific discoveries and ancient biblical wisdom.

Thanks to Noah and Yaakov Weinberg, Motty Berger, Ari Kahn, Gershon Unger, and Sam Veffer, all of whom helped me to discover depths of Torah. And thanks to those who shared their thoughts with me: Samuel Messinger, Mark Stuckey, Ruth Kolani, Celeste Sosnovik, Mel Weiner, Terry Esther Meeuwsen, Geof Greenfield, Jan Willem van der Hoeven, Aryeh Gallin, Peggy Ketz, Rita Greenfield, Philip Houseman, Paul Davies, Dennis Turner, Leon Lederman, Steven Weinberg, Tabitha Hofland, Rudolf Steinberg, Howard Sterling, Avraham Rosenthal, David and Daniel Lapin, Angela Sandler.

And most obviously, thanks to my wife, Barbara Sofer Schroeder, and our children, Avraham, Joshua, Hadas, Yael, and Hanna, with whom conversations turned into fountains of inspiration. And to my parents and my in-laws the Kahns.

As Whitney Houston acknowledged on the cover to her CD of *The Bodyguard,* "Thank you Heavenly Father for the strength and the love that only you could give."